Understanding Organisational Culture in the Construction Industry

T0186740

Spon research

publishes a stream of advanced books for built environment researchers and professionals from one of the world's leading publishers. The ISSN for the Spon Research programme is ISSN 1940-7653 and the ISSN for the Spon Research E-book programme is ISSN 1940-8005

Free-Standing Tension Structures: From tensegrity systems to cable-strut systems
978-0-415-33595-9
W.B. Bing

Performance-Based Optimization of Structures: Theory and applications
978-0-415-33594-2
Q.Q. Liang

Microstructure of Smectite Clays & Engineering Performance
978-0-415-36863-6
R. Pusch & R. Yong

Procurement in the Construction Industry: The impact and cost of alternative market and supply processes
978-0-415-39560-1
W. Hughes, P. Hillebrandt, D. Greenwood and W. Kwawu

Communication in Construction Teams
978-0-415-36619-9
S. Emmitt & C. Gorse

Concurrent Engineering in Construction Projects
978-0-415-39488-8
C. Anumba, J. Kamara & A.-F. Cutting-Decelle

People and Culture in Construction
978-0-415-34870-6
A. Dainty, S. Green & B. Bagilhole

Very Large Floating Structures
978-0-415-41953-6
C.M. Wang, E. Watanabe & T. Utsunomiya

Tropical Urban Heat Islands: Climate, Buildings and Greenery
978-0-415-41104-2
N.H. Wong & C. Yu

Innovation in Small Construction Firms
978-0-415-39390-4
P. Barrett, M. Sexton & A. Lee

Construction Supply Chain Economics
978-0-415-40971-1
K. London

Employee Resourcing in the Construction Industry
978-0-415-37163-6
A. Raiden, A. Dainty & R. Neale

Managing Knowledge in the Construction Industry
978-0-415-46344-7
A. Styhre

Collaborative Information Management in Construction
978-0-415-48422-0
G. Shen, A. Baldwin and P. Brandon

Containment of High Level Radioactive and Hazardous Solid Wastes with Clay Barriers
978-0-415-45820-7
R.N. Yong, R. Pusch and M. Nakano

Technology, Design and Process Innovation in the Built Environment
978-0-415-46288-4
P. Newton, K. Hampson and R. Drogemuller

Performance Improvement in Construction Management
978-0-415-54598-3
B. Atkin and J. Borgbrant

Relational Contracting for Construction Excellence: Principles, Practices and Partnering
978-0-415-46669-1
A. Chan, D. Chan and J. Yeung

Understanding Organisational Culture in the Construction Industry

Vaughan Coffey

Routledge
Taylor & Francis Group

LONDON AND NEW YORK

First published 2010
by Spon Press
2 Park Square, Milton Park, Abingdon, Oxfordshire OX14 4RN

Simultaneously published in the USA and Canada
by Spon Press
711 Third Avenue, New York, NY 10017

First issued in paperback 2014

*Spon Press is an imprint of the Taylor & Francis Group, an informa
business*

Typeset in Sabon by Glyph International Ltd

This publication presents material of a broad scope and applicability.
Despite stringent efforts by all concerned in the publishing process,
some typographical or editorial errors may occur, and readers are
encouraged to bring these to our attention where they represent
errors of substance. The publisher and author disclaim any liability,
in whole or in part, arising from information contained in this
publication. The reader is urged to consult with an appropriate
licensed professional prior to taking any action or making any
interpretation that is within the realm of a licensed professional
practice.

British Library Cataloguing in Publication Data
A catalogue record for this book is available from the British Library

Library of Congress Cataloging-in-Publication Data
Coffey, Vaughan.
    Understanding organisational culture in the construction industry /
Vaughan Coffey.
       p. cm.
    Includes bibliographical references and index.
    1. Construction industry. 2. Corporate culture.
    3. Success in business. I. Title.
    HD9715.A2C585 2010
    302.3′5–dc22   2009031574

ISBN 13: 978-1-138-86134-3 (pbk)
ISBN 13: 978-0-415-42594-0 (hbk)

Contents

Figures

Tables

Foreword

The overriding reason for carrying out the research described in this book was originally in fulfilment of my own university doctoral programme requirements. However, interest in the whole area of organisational studies originally emanated from my observation of problems that originated in my working field of over 20 years (i.e. the construction industry), in particular in the question of why some construction companies performed well consistently and others only sporadically, or not at all. I pondered the questions: What is different about the best performers and what gives rise to their success? This interest extended to closely observing the different management styles in action and the kind of organisational 'feel' that different companies possessed and it was what triggered me to start to read the extant literature about organisational and other cultural studies, as well as the large body of work that has been undertaken by those who have gone before me investigating organisational effectiveness and business success. While reviewing the literature in great detail, I soon came to the various discussions on the difference between organisational culture and climate, the arguments about which research methodology was best to study the concept of organisational culture and the dialogues that abound on the way to measure organisational effectiveness. Throughout the literature, I kept coming across the various studies on the perceived link between organisational culture and performance, and one name cropped up with regularity: Dr Daniel Denison. I began my studies by reading his book entitled *Corporate Culture and Organizational Effectiveness* and, from that point, my research ideas began to develop.

My studies have borne fruit in that I now have a fairly deep understanding of what it is within organisations that makes a successful or unsuccessful company. Indeed, the results reported by Denison and others have been closely replicated in my own work, thus giving further support to the premise that there is a relationship between 'strong' organisational cultures and effective business performance, that the Denison Organisational Culture Survey is useful in other ethnic and organisational cultures from those in which it was first developed, and that there are other ways to measure effectiveness rather than the previous financial or the more recent score-card models.

Once my doctoral thesis was successfully completed, I realised that any serious study of organisational culture carried out within the construction industry first required the investigator to fully understand the concepts and complex paradigms that were being dealt with. It was this realisation that persuaded me to set my work down for a wider audience in a book format, and I hope this is what this humble volume achieves.

Vaughan Coffey
Queensland University of Technology, Australia

1   An introduction to organisations, culture, performance and construction

Chapter introduction

This book, as the title describes, sets out to provide readers with an understanding of how organisational culture impacts on, and influences companies operating in the construction industry. The objective of this opening chapter is to present sufficient background for the reader to comprehend the purpose, scope and significance of the research and outcomes represented in this book. The chapter introduces the field of study, details the main questions underpinning the research described in later chapters of this book, overviews the research design and sets out the limitations of the research.

Background to the research

Organisational research[1]

Research into the areas of organisational culture and effectiveness, both as standalone studies as well as studies into the linkage between the two, originate from the field of organisational behavioural research. The following definition demonstrates well the foundation for the evolution of this branch of organisational study.

> Organisational behaviour (frequently abbreviated as OB) is a field of study that investigates the impact that individuals, groups, and structure have on behaviour within organisations, for the purpose of applying such knowledge toward improving an organisation's effectiveness.
>
> (Robbins, 1990: 8)

Most OB textbooks state that the study of organisational behaviour draws upon contributions from the fields of psychology, sociology, social psychology, anthropology and political science, and it is perhaps because of these multidisciplinary influences impacting on its study, together with the abundance of contingency variables that often occur in OB research that ensure there are very few universal principles to explain the complex questions that arise when delving into this area of organisational research and the debates ensuing on how to provide convincing answers to such questions (Tjosvold, 1987). This aside, much research has been undertaken into specific areas of OB such as organisational culture and organisational effectiveness, and the main themes of the book are set in those specific areas.

The organisational culture and performance link

Several researchers and practitioners have established evidence that a link exists between levels of development and prevalence of various organisational traits and the resultant levels of success of companies to deliver high-quality products and services and maintain market sector dominance (Denison, 1990; Gordon and DiTomaso, 1992; Kotter and Heskett, 1992; Petty *et al.*, 1995; Wilderom and van den Berg, 1998). Several other critical reviews of these research studies challenge the core hypotheses of this presumed link between organisational culture and performance (Saffold, 1988; Siehl and Martin, 1990; Lim, 1995; Wilderom *et al.*, 2000). The latter provide a thorough review of 10 major empirical culture–performance (C–P) link studies carried out in the United States and Europe, and conclude: 'most of the empirical studies reported in the 1990s had important conceptual and methodological weaknesses' (2000: 196). The conflict that exists in this area of research is evident even among the most severe critics of the C–P link; take the case of Lim who states (1995: 20) that 'as it stands, the present examination does not seem to indicate a relationship between culture and the short-term performance of organisations, much less show a causal relationship between culture and performance' and yet concedes in a later study that (2000: 2) 'the most important contribution of "culture" towards the understanding of organisations appears to be as a descriptive and explanatory tool, rather than a predictive one'. Similarly, Wilderom *et al.* (2000: 201), who were critical to a degree of the previous studies, conclude that 'nevertheless, the great intuitive appeal of the C–P linkage, the preliminary evidence found so far and the many research challenges involved in obtaining the evidence give some reason to still believe in this link'. Various other authors have sought to follow a different research route and examine and measure the 'culture gap' that exists between the perceived and preferred states that exist within organisations (Wilderom and van den Berg, 1998; O'Reilly *et al.*, 1991), although the empirical evidence from these studies has not so far provided any strong proof of a C–P linkage.

Measuring organisational culture

Researchers using quantitative methods to investigate the construct of organisational culture have found difficulty in actually defining and measuring the construct, mainly as a result of the wide range of operational definitions of the term, but also owing to varying perceptions of the concept of organisational culture used in different studies. Another difficulty faced by researchers arises from the lack of agreement on an acceptable universal framework for comparing different cultures that exist even within similar organisations. Wilderom *et al.* (2000: 201–202) raise this critical issue in their review mentioned earlier and question whether any of the studies they examined actually measure the construct of organisational culture. Their view on the 10 studies examined is summarised in Table 1.1.

An ongoing debate has been occurring in recent years among various contemporary researchers about the efficacy of using either the 'culture' or 'climate' approach when assessing organisational metrics and, from a thorough review of the literature carried out by the author, it would appear that the major difference between the two approaches is that the former analyzes typical organisational practices that produce measurable outcomes and the latter examines the overt views which managers and employees hold about their organisations at any given point in time. Denison, commenting on this debate (1996), concludes that researchers should not waste their efforts participating in a 'paradigm war' and that it would be better for climate-based research to support and amplify culture-based research, in order to provide a richer overall picture of the different organisational cultures that exist within companies and then relate these to the business outcomes of such companies. This approach to the assessment of organisational culture is examined in more detail later in the book and a detailed explanation will be given as to why the research design described by the author in his study of the Hong Kong construction industry

Table 1.1 Organisational attributes/dimensions measured by recent C–P studies

Reference	*Date*	*Organisational dimensions/attributes measured*
Calori and Sarnin	1991	Structural aspects of an organisation
Marcoulides and Heck	1993	and work-related values
Rousseau	1990	Normative beliefs
Denison	1990	Organisational climate
Gordon and DiTomaso	1992	
Denison and Mishra	1995	
Kotter and Heskett	1992	General and practical functioning
Petty *et al.*	1995	of organisations
Koene	1996	

Source: Wilderom *et al.* (2000: 201).

draws from a framework derived from the view put forward by Sparrow (2001: 88) that:

> both culture and climate assessments describe ways in which individuals make sense of their organisation and provide the context for organisational behaviour, allowing researchers, practitioners or consultants to describe, explain or even predict why some behaviours are more effective than others for a particular organisation.

The instrument used to measure the construct of organisational culture in the research described in this book is based on the *Denison Organisational Culture Survey* (DOCS) originally developed from a concept originating from an earlier study (Denison and Mishra, 1995) and this is described in detail in Chapter 5. A diagram adapted by the author and derived from the current Denison (2009) model is shown in Figure 1.1.

Measuring performance and effectiveness

Another fairly contentious issue in forging the link between organisational culture and effectiveness relates to the choice of what measures of a

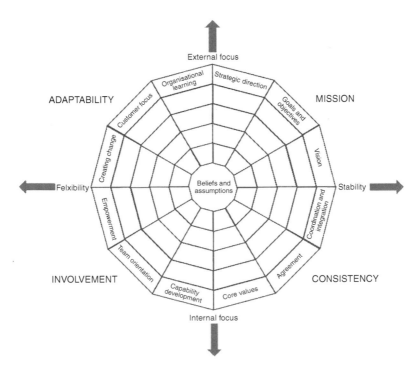

Figure 1.1 The Denison organisational culture model.
Source: Denison (2009) adapted by Coffey.

company's performance (or of an organisation's effectiveness) are actually used and how such measures are then compared to establish what reflects better (or best) and worst performance in any particular sector. Most of the studies carried out between the mid-1960s and the early 1980s used easily accessible rational goal performance measures such as accounting-based indicators (e.g. profitability, return on equity, return on sales, return on investment and so on). Such measures were later considered by some authors and researchers to be unsuitable due to increasing evidence that the perception of organisational effectiveness could be distorted if inadequate or inappropriate goals were selected (Miles and Ezzell, 1980; Mohr, 1983). Thus, goal-based benchmarks were eventually superseded by systems-based models; however, a view developed once again that although systems-based measures focused on a much broader set of variables, they too could be incorrectly set so as to cause unwanted external consequences (Mohr, 1983). As a result of these concerns regarding the appropriateness of previous measures, new research models were developed. Rojas (2000) describes these as 'multiple-constituency models' which as well as measuring effectiveness internally also measured it as a function of overall customer satisfaction. Quinn and Rohrbaugh (1983) attempted to classify how various management experts, practitioners and other researchers actually described organisations based on the statistical factor analysis of a comprehensive list of effectiveness indicators. As a result of their research they developed the Competing Values Framework (CVF) model which showed that organisational theorists tended to share a cognitive map, or an implicit theoretical framework of organisational effectiveness. As there is a developmental relationship between the concept of competing values and the development of DOCS, the CVF will be discussed later in the book when the metrics of organisational effectiveness are covered. There were several variants of the original CVF model that were developed during the 1990s and which have been used by other researchers to examine effectiveness (Bhargava and Sinha, 1992; Ridley and Mendoza, 1993; Jackson, 1999). However, despite the large amount of research undertaken with the model, there is still little consensus on a valid universal set of criteria for measuring effectiveness. By the early 1990s, a growing body of opinion had emerged that opposed these kinds of measures (Kaplan and Norton, 1992; Brown and Laverick, 1994; Baltes, 1997) and, while goal-based measures of organisational effectiveness were largely felt to be defunct by many authors by the end of the 1980s, they continued to drive the measurement and assessment of corporate performance well into the mid-1990s. In order both to avoid such criticism and to bring a new dimension to the area of study, the research described in this book uses a non-financially based and yet highly objective set of metrics (i.e. the Hong Kong Housing Department (HKHD) Performance Assessment Scoring System (PASS)) to measure and compare the organisational performance of Hong Kong-based construction companies operating in the public sector housing arena. Performance monitoring in the Hong Kong construction industry and PASS are briefly described later in this chapter and in greater detail in Chapter 5.

The culture of construction companies

Little work has been undertaken globally on examining or measuring the culture of construction companies (Root, 1994), and this certainly holds true in South East Asia and particularly in Hong Kong, despite the construction industry in the latter accounting, during the period of the mid-1980s to the mid-2000s, for such a high percentage of GDP generally, and of government's expenditure in particular. Over the past 10 years or so, several ground-breaking reports have been produced in the United Kingdom, Europe and Australia as well as in Hong Kong, that have outlined the need for the construction industry to reorganise and change its existing culture, in order to guarantee better quality product, greater customer satisfaction and generally to become more competitive and effective. In addition, a Task Group of the International Council for Research and Innovation in Building and Construction (CIB TG23[2]) between 1993 and 2006 undertook research specifically into the role and impact of culture in the construction industry. The reports of, and the research findings from, TG23, together with the narrower range of literature dealing specifically with the subject of construction industry culture are reviewed in Chapters 3 and 5.

Measuring the performance of Hong Kong construction companies

Ever since the 1970s, the HKHD together with other works departments of the Hong Kong Special Administrative Region Government (HKSARG) operated a common 'contractors' performance review system', used primarily to support the HKSARG's Central Tender Board (CTB) and Hong Kong Housing Authority's (HKHA) Building Committee (BC), when both were considering and deciding on which companies to award building works contracts to. This earlier performance reporting system operated on a subjectively based 'tick-box' grading system ranging as follows:

A – Excellent
B – Very Good
C – Moderate
D – Poor
E – Very Bad

The system, while having the distinct advantages of being easy to operate and applicable across a variety of different types of construction project, was highly subjective in nature, meaning that much care was required to be exercised in using such metrics. At this juncture, in line with all other government works departments, the HKHA's building tenders were generally invited by publishing a gazette notice, and applications to tender were accepted on an 'open' basis from a large 'Approved List of Contractors for Public Works'. This system of tendering, in the view of the HKHA, had the disadvantage of bringing in often over-competitive 'cut-throat' bids from companies who were unknown to the HKHD, or often even to the other works departments. The HKHA as a publically funded entity felt this to be an increasingly high-risk scenario for taxpayers to bear, with a grave potential for poor performance or financial failure of companies that it employed. In view of such concerns, the HKHA in April 1990 promulgated its own 'List of Building Contractors'. The use of a specific list of contractors and direct invitation to tender for HKHA projects was a radical departure from the previous system operated government-wide. The new list obviously required the implementation of new 'tools' to assist in managing it. The development and implementation of a new objective performance monitoring system to replace the former subjective 'tick-box' system was therefore considered to be of paramount importance. The Performance Assessment Scoring System (PASS) was developed to determine objectively the performance of the construction companies employed by the HKHA and is based on measuring the construction quality produced by them on projects undertaken for the HKHA.

PASS – an objective measurement tool

The earliest version of PASS was developed by a team of construction professionals from the HKHD's Construction Branch in late 1990. Since its promulgation that year and full implementation in 1991, the PASS has been used to objectively measure the performance of building contractors directly against well-defined standards. The system has also provided a fair means of comparing the performance of individual companies and rewarding those best performers with superior tendering opportunities. Not only does PASS enable detailed and regular objective performance reporting to be provided to HKHD on all of its projects and their contractors, but the Department's List Management Committee (LMC) has been able to use the scores emanating from the overall performance reports generated from PASS when awarding preferential tendering opportunities. HKHD's project teams and consultants use PASS as a systematic quality checklist and the building contractors use it as a key performance and quality improvement indicator as well as a process control tool. The system has been developed and refined over the years and has proven to be a highly useful tool for selecting better performing contractors to tender for upcoming projects. The incentive of a larger number of tendering opportunities allocated for better on-site performance has been an important catalyst in

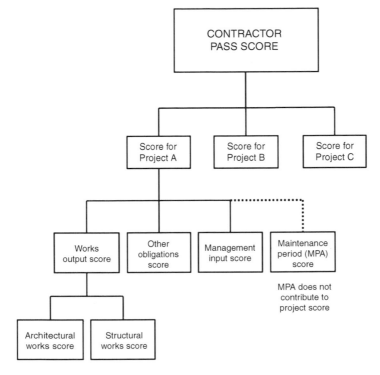

Figure 1.2 The simplified Performance Assessment Scoring System (PASS) model.
Source: HKHD (2000) adapted by Coffey.

improving on-site standards of work and operations of many of the listed contractors.

A detailed description of the development and operation of PASS and the nature of data and information produced from the system is given in Chapter 5; however, it is useful to the reader at this stage to give a brief summary of the basic concept and nature of PASS, as part of the background description to this research. Figure 1.2 shows a simplified conceptual model of PASS.

The overall performance improvement of contractors on projects has in part been due to the reward component of the system, but also to the disciplinary actions imposed on poor performers. These actions range from a three-month restriction from tendering through to longer periods of formal suspension, and in extreme cases, removal from the list. With the exception of the removal in 1993 of 63 companies due to their not attaining the required ISO 9001:1987[3] certification by the official deadline date set by the HKHD, the actual number of removals from the HKHA list and suspensions for disciplinary reasons of companies on the list has been quite small. However, the numbers of contractors restricted from tendering for a prescribed three-month interval due to poor PASS results since 1997 has been more significant. Given the existence of quality problems described in a later section of this chapter that directly affected both the business success of companies involved in the construction of public housing as well as the customer satisfaction of recipients of their constructed product (i.e. HKHD and the tenants of public housing), the possession and maintenance of a robust and accurate performance monitoring system is one of the most important assets available to the HKHD's construction managers and project teams.

Research objectives, problems and hypotheses

Objectives of the study

Based on the above background issues and in particular the core issue of investigating the existence of a demonstrable link between organisational culture and performance and supported by the theories outlined in the literature review which follows, two major objectives for the research described in this book were defined as follows:

- Measure internal and external perceptions of four traits of the determinants of organisational culture (i.e. adaptability, involvement, consistency and mission) within listed construction companies.
- Test the theory of whether Hong Kong construction companies engaged in public housing projects perform more effectively if they exhibit relatively high levels of these four organisational cultural 'traits'.

Research problem

Contemporary literature and academic sources on construction management have tended to concentrate on the methodologies necessary for planning and executing construction projects rather than on the organisational cultures and philosophies which delineate successful construction companies from unsuccessful ones. Chinowsky and Meredith (2000) have pointed out that unlike other types of manufacturing industries (e.g. automotive, personal computers) that produce large numbers of similar products, the output from normal construction projects is a unique and specific 'end product'. The interest of construction company management is, not unsurprisingly, therefore focused on the 'project format', i.e. the planning, control and execution of projects within their own frameworks (Halpin and Woodhead, 1998). Because of this, the research problem that develops out of the combination of the historical scenario previously outlined and the lack of research specifically directed at critically examining construction company organisational cultures, values and frameworks is one of determining the attributes and traits within such companies that are most significant in contributing to the successful quality outcomes on public housing building projects. As has been mentioned, earlier research on organisational culture usually measured financial or quantitative output metrics to determine effectiveness, and so the use of PASS to determine business success (organisational effectiveness and performance) brings something entirely new to the specific area of culture performance research.

Research questions

Based on the research objectives and main research problem described previously, the resultant research questions that emerged for the author to pursue (and that are also covered in detail later in the book) are:

- *Question 1* – Do Hong Kong construction companies possessing relatively high combined levels of the four organisational cultural 'traits' (i.e. adaptability, involvement, consistency and mission) (as indicated by the Denison Organisational Culture Model) perform more successfully on public housing projects than those exhibiting lower levels of those traits?
- *Question 2* – Are any of the four traits more significant in contributing to success levels than others?
- *Question 3* – Are any combinations of the four traits, based on a horizontal or vertical split of the Denison Organisational Culture Model, more significant in contributing to success levels than others?

Figure 1.3 illustrates the theoretical framework of the research questions.

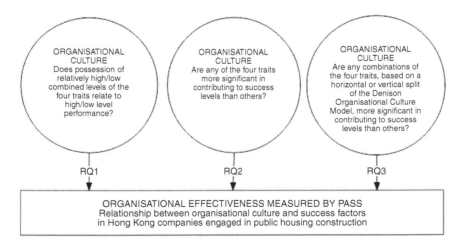

Figure 1.3 Theoretical framework of research questions on the link between organisational culture and effectiveness.

These research questions and their stated objectives translate into operational statements (hypotheses), which are described in more detail in Chapter 6.

Justification for the research

Whenever issues concerning public housing in Hong Kong are discussed, be it by the media, in public debate or within an academic research context, it would be very difficult for anyone to understand the importance of such issues (unless they were closely involved) without some basic background as to the significance of such issues within the overall 'big picture' of Hong Kong. What follows is a brief description of the historical background of public housing development in Hong Kong over the past 45 years and a setting of the subject within a very specific research arena.

Public housing in Hong Kong (1953 to 1997) – a brief history

Walker (1996) has noted that prior to 1841 (i.e. the year that the British arrived in Hong Kong), there were around 5,000 people living in walled villages on the mainland north of Hong Kong (now known as the New Territories) and some 7,000 living on boats in the bays and typhoon shelters immediately offshore. Independent building companies that originated in, and after, 1841 were usually individuals or groups of people offering trades and services and working under architects and engineers, or in the case of government work, the Surveyor-General. Following the ceding of Hong Kong to Britain in 1841 (during the First Opium War), this scenario prevailed, and growth of population in Hong Kong increased steadily over the next 100 years. This all began to change drastically in 1926 when war broke out between the Communist Party (led by Mao-tze Tung) and the Nationalist Kuonmintang (led by General Chiang Kai-shek) in China. One author summarises this turbulent period thus:

> Between 1947 and 1949, in the midst of the civil war on the Chinese mainland, masses of refugees surged into the territory. The better off put their capital into the construction of squatter huts for rental profit. The population grew larger, squatter huts grew denser.
>
> (Leung, 1999: 23)

By 1949, the population of Hong Kong had grown dramatically to over two million people, around 300,000 of whom lived in the squatter areas. On 25 December 1953, a serious fire broke out in a particularly densely populated squatter village at Shek Kip Mei, which was situated on the Kowloon Peninsula. Neither the residents nor the fire brigade were able to put the fire out and it continued to burn until the following day. When a count was made, around 50,000 people had lost their homes and, of these, 20,000 had nowhere to go and were forced to erect their ramshackle and burned-out huts on the streets to act as temporary shelter. The government, in order to accommodate the victims, ordered the Public Works Department to construct temporary housing shelters immediately (which they managed to do in a matter of two months) and the Urban Council set up an emergency committee to look at a longer term solution to the problem of housing the 'squatters'. The former Resettlement Department and the Housing Section of the Urban Services were merged in 1972 to form the new Housing Department (HD), which became the executive arm of the Hong Kong Housing Authority (HKHA) and was given the responsibility for developing new public housing designs and arranging and overseeing the public housing programme building works. The last Mid-year Population figures supplied by the Demographic Statistics Section of Census and Statistics Department (C&SD) show that at the end of 2008, the population of Hong Kong stood at just over seven million people and that approximately 48 per cent are housed in public rental or home-ownership flats. On 1 July 1997, the sovereignty of Hong Kong was transferred back from the British government to the Peoples Republic of China, and in his inaugural 1997 Annual Policy Address, the newly created Hong Kong Special Administrative Region's (HKSAR) first Chief Executive (CE), Mr Tung Chee-hwa, announced a 10-year housing programme under which he pledged 85,000 flats per year to be built before 2006–2007, of which 50,000 each year were to be for rental or home ownership. It was anticipated under this plan that some 70 per cent of Hong Kong people would own their own homes by 2007 and some 2.2 million people would be living in some form of public housing by that time.

Public housing, 1997 to the present day: a current status report

By 1 April 1998, when the CE's new programme was less than one year into operation, it had become clear that an annual flat production figure of 85,000 was maybe an impossible target to achieve. A number of factors had arisen that had a significant impact on the programme and its objectives. These were as follows:

- The declaration by private residential developers of their inability to produce their allocated percentage of the proposed output, due to a 'flattened' property market.
- Acute labour shortages in certain crucial sectors of the construction industry, caused by relocation of workers into more attractive and well-paid sectors of the manufacturing industry.
- Practical difficulties experienced by the construction industry in shortening construction periods to achieve the required output levels without allowing deterioration in the construction quality.

- The introduction of more restrictive legislation covering the operation of construction sites and of construction companies themselves.

Nevertheless, the actual public housing flat production totals achieved against programme by the HKHA between 1997 and 2008 since promulgation of the original output targets (from 1 April to 31 March of each year shown) were impressive (particularly from 1998 to 2002 – the period of the author's original research). These production figures are shown in Table 1.2.

Quality problems exposed – 1999/2000

While the HKHA was no stranger to the experience of having construction quality problems exposed (in the 1980s 26 blocks of flats had been discovered to have serious structural problems requiring immediate demolition or substantial repair and by 2000 all had been demolished), it is probably fair to say that the particular problems exposed in the 1999/2000 period came as a surprise in terms of both the numbers and their magnitude. They may be divided into two main categories, namely those discovered by internal auditing exercises and those resulting from external events.

1   *Internal discovery*: The HKHD's Technical Audit Unit (TAU), which carries out periodic reviews of construction operations and standards (ensuring their compliance with client specifications and acceptable trade practices) on HKHA projects, discovered serious non-compliances in finishing works on 17 recently completed projects. These discoveries

Table 1.2 HKHA production/sales (1998–2008)

Year	Total housing unit production
1998/1999	28,499
1999/2000	48,484
2000/2001	89,002
2001/2002	39,405
2002/2003	20,390
2003/2004	15,468
2004/2005	24,682
2005/2006	17,153
2006/2007	8,392
2007/2008	15,112

Note: Figures shown are aggregated from completion of subsidised rental and private sales completions of HKHA units.

Source: HKHA (2000).

resulted in a review of specifications and standards, overall programme requirements and construction periods, and contract management and site supervision procedures. A detailed review of the existing contractors' performance monitoring and reporting systems was undertaken and a report was published which resulted in the re-engineering of certain components of both the system and assessment methodologies. As these improvements were not implemented until January 2002 (due to a prolonged period of consultation and negotiation with the construction industry concerning some of the features of the proposed new system) the PASS 1997 version continued in use in a 'stable state' until that time, and it was this version that was used to generate the data for the research described in this book.

2   *External discovery*: At the beginning of 2000, the newspapers in Hong Kong carried banner headlines that alluded to the discovery of inadequate length and insufficient numbers of pre-cast concrete foundation piles beneath one of the new stations being constructed for the Mass Transit Railway Corporation (MTRC)'s Airport Express line. This discovery prompted the instigation of investigations by various HKSARG 'building' and 'civil' works departments that had projects being undertaken by the main contractor and subcontractors employed on the affected MTRC project. In the case of the HKHD, several foundation contracts were under construction by the same contractor and/or subcontractors. Rigorous structural investigations by an HD special task force revealed that, in addition to these projects, a further three were affected to varying degrees of magnitude by 'short' or insufficient piling deficiencies. In the most serious case, 50 per cent of piles were under length. In order to tackle the perceived root causes of this and the other problems mentioned earlier (i.e. of poor finishes to flats and public housing construction quality issues in general), the HKHA embarked on a study of existing practices and procedures, and the outcome of this review was the publication of a consultative document entitled '*Quality Housing: Partnering for Change* (HKHA, 2000). The document, which contained 40 (which later increased after consultation to 50) recommendations for changes in the operational and organisational methodologies of the HD and sought much greater partnered team working, directed at producing improvements in quality outcomes, between its own professional and site supervisory staff and the management teams of its contractors. Details of recommendations from this report specifically targeted at construction companies together with recommendations from an HKSAR government-driven report entitled *Construct for Excellence* (CIRC, 2001), which clearly stated the perceived need to change the 'culture of the industry' if construction quality in Hong Kong was to improve. A major part of the public and media attention focused on why the various performance monitoring and quality control systems known to be operated by the HKHD had apparently failed.

This then sets the scene within which the population for the research described in this book, namely the Building (New Works) Category construction companies on the HKHA's list, are operating, justifies the need for investigating those aspects of company culture which contribute to successful quality and client-measured success outcomes and also helps explain why the efficacy, accuracy and objectiveness of the HKHD's performance monitoring system constitute an important aspect of the specific research described.

Research methodology

The paradigm of the research and the main methodologies used in the later chapters of this book that describe the specific research of the author into the linkage between organisational culture and performance are described in detail in Chapter 5, but may be summarised as follows:

Research paradigm and methodologies used in this book

Set in the positivist paradigm, the main methodology uses primary data gathered from the DOCS undertaken in a representative sample of the specific research population, i.e. public housing construction companies in Hong Kong. The results of this survey are used to develop a metric that informs the concept of organisational culture and allows measurement of its comparative strength in each company surveyed. These data are then compared with secondary data of measured performance levels, (i.e. PASS results) of the same companies to ascertain whether the culture–performance link exists and also whether Denison's various hypotheses concerning such linkage can be supported in the specific area of interest in the local Hong Kong scenario, The original Denison instrument was adapted using the established procedures of decentring and back-translation for use in a different national (ethnic) culture from that in which it was originally developed and operated. Detailed quantitative analysis involving correlation and regression tests, as well as some preliminary qualitative study, in relation to the value concepts of participants and self-performance rating of company performance, was also carried out and the results analysed and used to assist in triangulating the data and findings for greater rigour in developing conclusions.

Limitations of scope of the research and key assumptions

Certain limitations were identified in the research and these may be summarised as follows:

- Only companies on the Hong Kong Housing Authority (HKHA)'s List of Building Contractors, Building (New Works) Category, Groups NW1 and NW2 have been used as the survey population.

- The numerical scores used to determine comparative success ratings of companies have been drawn from the Hong Kong Housing Department (HKHD)'s PASS 1997 edition and cover performance for the period from April 1997 to March 2000.
- Only building contracts undertaken for the HKHA have been used to provide performance.
- PASS data used are based solely on the 'output' (workmanship), 'progress' (against contract programme) aspects of performance and on the 'input' (management) aspects.

Chapter summary

This chapter has provided sufficient background for the reader to comprehend the purpose, scope and significance of the research and outcomes represented in this book. The chapter introduced the overall field of study, detailed the main questions underpinning the research described in later chapters of this book and overviewed the research design within the context of certain limitations.

Notes

1　The reader will note that in the main authored text, the spelling of *organisation* will be in the UK/Australian English form, but where citations, quotations or sources from the US are used, it will be spelt as *organization* in the US English form.
2　Since 2006, former CIB TG23 has become Working Commission W112 on *Culture in Construction* and will continue with an extended scope from the work done by the former.
3　ISO 9001:1987 Quality systems – Model for quality assurance in design/ development, production, installation and servicing (International Organisation for Standardization: Geneva).

2 Organisations, culture and climate

Chapter introduction

Before studying the concept of organisational culture in any particular sector of industry, the reader needs to acquire a thorough knowledge of the nature and development of organisational study and research, and in particular to examine the interest in and development of the study of organisational culture in a range of organisations. It is important to be able to answer questions such as 'How did such an interest develop?', 'Who were the main contributors in the field?' and 'What impact did this field of study have on modern organisations?' Chapter 2 reviews and analyses the relevant literature that has appeared on this issue over the years and sets the scene for subsequent chapters that examine closely the literature on the relationship between 'organisational culture' and 'effectiveness'.

What is an organisation?

In order for the reader to understand specific organisational facets such as culture or effectiveness, there is a need to examine a more basic organisational concept, and so it is useful to define exactly what an *organisation* is. Barnard (1938: 73) classically defined an organisation as 'a system of consciously coordinated activities or forces of two or more persons'. In a more recent text, Kreitner and Kinicki (2001) give this 'conscious coordination' aspect four characteristics, which are as follows:

- hierarchy of authority;
- coordination of effort;
- common goal;
- division of labour.

Daft (2001: 12) acknowledges the diversity of organisations but usefully leads the reader through an additive definition: 'organisations are (1) social entities that (2) are goal directed, (3) are designed as deliberately structured

Table 2.1 Definitions of organisations

'A social relationship, which is either closed or limits the admission of outsiders by rules, will be called 'corporate group' so far as its order is enforced by the action of specific individuals whose regular function this is, of a chief or 'head' and usually also of an administrative staff' (Weber, 1913/1947: 145–146).

An organisation is:

1   a plurality of parts
2   maintaining themselves through their interrelatedness, and
3   achieving specific objectives,
4   while accomplishing (2) and (3) adapt to the external environment, thereby,
5   maintaining their interrelated state of the parts (Argyris, 1960: 27–28).

'[Organisations are] ... structures of mutual expectation, attached to roles which define what each of its members shall expect from others and from himself' (Vickers, 1967: 109–110).

'Organizing ... is defined as a *consensually validated grammar for reducing equivocality by means of sensible interlocked behaviors*. ... Organizations keep people busy, occasionally entertain them, give them a variety of experiences, keep them off the streets, provide pretexts for story-telling, and allow socializing. They haven't anything else to give' (Weick, 1979: 3, 264).

'[An organisation is] ... collectivity with a relatively identifiable boundary, a normative order, ranks of authority, communications systems, and membership-coordinating systems; this collectivity exists on a relatively continuous basis in an environment and engages in activities that are usually related to a set of goals; the activities have outcomes for organisational members, the organisation itself, and for society' (Hall, 1987).

'[Organisations are] ... social arrangements for the controlled performance of collective goals' (Huczynski and Buchanan, 1991).

'Organisations are nets of collective action, undertaken in an effort to shape the world and human lives. The contents of the action are meanings and things (artifacts). One net of collective action is distinguishable from another by the kind of meanings and products socially attributed to a given organisation' (Czarniawska-Joerges, 1992: 32).

'[An organisation is] ... a consciously coordinated social unit, composed of two or more people, that functions on a relatively continuous basis to achieve a common goal or set of goals' (Robbins, 1998: 2).

Source: Gabriel and Schwartz (1999) adapted by Coffey (2006).

and coordinated activity systems, and (4) are linked to the external environment.' Gabriel and Schwartz (1999: 58) cite several definitions of the term *organisation* summarised in Table 2.1.

Drawing from this table, there appear to be some common factors emanating from these otherwise diverse definitions, namely that organisations have established boundaries and defined social structures within which all the participants have roles and take coordinated actions; a communication system exists; there are goals set and outcomes produced, and the whole sits within a larger universal environment. As Gabriel and Schwartz (1999: 58) point out, organisations can be treated as 'facts' (i.e. they exist as factual and complete realities); however, the authors also note that despite over a hundred years of research into organisations, 'scholars have failed to come up with a satisfactory and generally acceptable definition of an organisation'. Gabriel and Schwartz also observe (1999: 59) that based on such diverse definitions and opinions, 'some scholars of organisations (for example, Weick, 1979; Sims, Fineman and Gabriel, 1993; Hatch, 1997) avoid definitions of organisations altogether and content themselves with loose descriptions of the meanings assumed by the term "organisation".' Swanson (1976: 665) highlights this dilemma more critically, describing it as follows: 'The recurrent problem in sociology is to conceive of a corporate organisation, and to study it, in ways that do not anthropomorphize it and do not reduce it to the behavior of individuals or of human aggregates.'

The critical question of interest arising from this discussion becomes: Why is it that the descriptions, definitions and perspectives of organisations appear to differ so much and 'what' definition of an organisation could relevantly be used in the context of this book? The main reason for the differences highlighted above appears to derive from the fact that organisational theory and design have varied over time, both in response to changes in society at large and (within the business context) to changes in management thinking. Scott (1998: 3) alludes to the fact that the older civilisations such as those of China, Greece and India harboured a variety of organisations, involved in the primary activities of 'soldiering', 'tax-collection' and 'public administration', and it was only in the industrialised societies of the late nineteenth and early twentieth centuries that the phenomenon of 'large numbers of organisations engaged in performing highly diverse tasks' evolved. Scott (1998: 4) continues by stating that to those primary historical activities of such societies were then added a variety new tasks ranging from 'discovery' (e.g. research organisations) and 'service providers' (e.g. laundries, medical care facilities), to 'communication bodies' (e.g. TV/radio, telephone companies). Now almost taken for granted, such organisations developed historically in a non-controversial, low-profile manner, but as one author remarks, 'the development of organisations is the principal mechanism by which, in a highly differentiated society, it is possible to "get things done," to achieve goals beyond the reach of the individual' (Parsons, 1960: 41). So if organisations were changing over time in this way, the question arises: What were the major shifts in theoretical thinking that were driving these changes? Table 2.2 briefly summarises the historical development of major organisational theories over the past 100 years.

Closer examination of this table gives some clues as to why the approach to organisational research (which provides the foundation from which more specific research studies into facets such as culture and effectiveness stem) has changed in the way that it has. The focus of organisational studies

Table 2.2 Brief historical overview of organisational theory

Organisational theory prior to 1900 – Emphasised the division of labour and the importance of machinery to facilitate labour.

Scientific management (1910s) – Described management as a science with employers having specific but different responsibilities; encouraged the scientific selection, training and development of workers, and the equal division of work between workers and management.

Classical school (1910s) – Listed the duties of a manager as planning, organising, commanding employees, coordinating activities and controlling performance; basic principles called for specialisation of work, unity of command, scalar chain of command and coordination of activities.

Human relations (1920s) – Focused on the importance of the attitudes and feelings of workers; informal roles and norms influenced performance.

Classical school revisited (1930s) – Re-emphasised the classical principles.

Group dynamics (1940s) – Encouraged individual participation in decision-making; noted the impact of work group on performance.

Bureaucracy (1940s) – Emphasised order, system, rationality, uniformity and consistency in management; led to equitable treatment for all employees by management.

Leadership (1950s) – Stressed the importance of groups having both social task leaders; differentiated between Theory X and Y management.

Decision theory (1960s) – Suggested that individuals 'satisfice' (a portmanteau of 'satisfy' and 'suffice') (i.e. obtain an outcome that is just good enough) when they make decisions.

Socio-technical school (1960s) – So-called for considering technology and work groups when understanding a work system.

Environmental and technological system (1960s) – Described the existence of mechanistic and organic structures and stated their effectiveness with specific sorts of environmental conditions and technological types.

Systems theory (1970s) – Represented organisations as open systems with inputs, transformations, outputs and feedback; systems strive for equilibrium and experience 'equifinality'.

Contingency theory (1980s) – Emphasised the fit between organisation processes and characteristics of the situation; called for fitting the organisation's structure to various contingencies.

Source: Wertheim (2001) adapted by Coffey (2005).

during the period from the turn of the century until the 1920s (Classical) was generally based on examining the efficiency of operatives in the workplace; the research methodologies used consisted of time-motion studies typified by the work of researchers such as Mary Parker Follett, Frederick Taylor and Frank and Lillian Gilbreth. By the time the century was approaching its mid-point, the behavioural psychologists had entered the research field and prominent within this period (Human Relations) were Fritz Roethlisberger and colleagues Elton Mayo, Chester Barnard and Kurt Lewin. The emphasis then shifted in research circles to a focus on turnover, absenteeism and worker well-being, and the major research methodologies were centred on employee satisfaction surveys. Contemporary organisational research is focused on aspects such as understanding people at work, creating effective work groups, leadership and management, and managing effective organisations. Specific research topics include organisational socialisation, individual and organisational learning, career development, work stress, interpersonal perception and attribution, group dynamics and self-managed work teams, inter-group conflict and negotiation, organisational culture, organisational design and organisational change. This change in organisational research focus mirrors the substantive changes in popular organisational theoretical perspectives, which have also taken place since the 1960s.

In relation to construction organisations which are the central focus of this book, two distinct contemporary organisational theories and the organisational structure which results from the application of these theories merit a closer look, namely 'systems' and 'contingency' theories. Systems theory surfaced in the 1960s following on from the interest in, and development of, management science and operations research during the 1950s immediately after the end of the Second World War. The interest that developed in new management styles as a response to wartime logistics and production problems was essentially a technological one and also contributed to the early interest in quality management systems which began to flourish at this time. According to Rainey (2003: 45): 'Through the 1960s and 1970s, an upsurge in empirical research on organisations extended and tested the open-systems and contingency-theory approaches and added new "contingencies" to the set.' One influential book authored by Woodward (1965) described a groundbreaking study carried out in the UK among three types of industrial firms: small batch or unit production systems companies (e.g. shipbuilding), large batch/mass production companies (e.g. automotive manufacturers) and continuous production companies (e.g. oil refineries). The major finding from this research, which has some bearing on this book, is that successful firms within the same category had managed to adapt their structure effectively to the demands of new and changing technologies and that successful firms within each specific category showed similar management structure profiles. It is interesting to note that when the structures of some successful construction companies in the Hong Kong construction industry are examined in Chapter 9 of this book, similar phenomena may be observed.

Leavitt (1965: 1145) was one of the earliest authors to describe a 'systems-based' view of organisations, which he based on four major internal components; task, people, technology and structure. Leavitt's fairly rigid 1965 model, which was derived to describe typical technological industry-based organisations of that time, also shows them as being responsible for 'processing' raw inputs into finished outputs under external environmental influences. It was not surprising that such a model should be developed by Leavitt, since it was derived from the prevailing perception of successful organisations being promoted by the 'quality gurus' that preceded him in the 1950s and 1960s, such as Deming, Feigenbaum, Crosby, Taguchi and others. Thus Leavitt's concept of an organisation bears a very strong resemblance to the 'simple process model' where inputs are transformed into products and services under the external influence of controls and based on the application of suitable resources. This process-driven concept of management developed with the quality movement in the 1980s and has often more recently been used in relation to describing quality management systems (ISO 9000 series) and total quality management (TQM), as well as in relation to the operations of building contractors when involved with building construction projects. A simple process diagram illustrating this concept is shown in Figure 2.1.

This model typifies in many ways how construction organisations operate and thus gives credibility when studying these types of organisations, to the use of research models that measure organisational input and the resource strength which is assumed to impact on it, and systems that measure inputs to, outputs from, and control of, the construction process. The systems theory perspective is highly relevant to the study of construction organisations, as central to the theory is the concept that any organisation when viewed as a whole comprises a set of subsystems and that there is a need to understand how all parts of a system interact with one another. The management and research interest then becomes the examination of the relationship between the organisation and its environment. What systems theory attempts to do is to comprehend the basic structure of any organisational system and from this deduce or investigate the behaviour among organisational participants that it produces. Senge (1984: 14) states: 'System dynamics is the study of complex systems, including such human systems as families, organizations, cities, and nations. If you look deeply into any system and analyze the relationships between members, you will

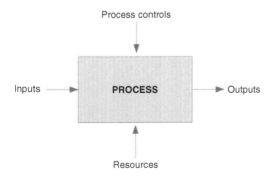

Figure 2.1 Simple process diagram.
Source: Coffey (2005).

find infinite complexity.' The analysis of an organisation using the systems approach would involve steps such as:

- defining the problem and the scope of system to be studied;
- breaking down the system into basic components;
- gathering data about each component;
- identifying/evaluating alternative solutions and selecting the best one;
- evaluating interactions among components of alternatives.

In construction organisations, this typifies the approach to engaging in, and mobilising for, projects, defining pricing strategies, establishing programmes and choosing subcontractors and partners. This makes it a highly suitable theoretical perspective from which to examine construction organisations. More recently, Scott (1998: 24–28) further developed this systems view of organisations and introduced three definitions of systems organisations built from the extant literature, which are summarised as follows:

- *Rational systems* – organisations are collectivities oriented to the pursuit of relatively specific goals and exhibiting relatively highly formalised social structures.
- *Natural systems* – organisations are collectivities whose participants are pursuing multiple interests, both disparate and common, but recognise the value of perpetuating the organisation as an important resource. The informal structure of relations that develops among participants provides a more informative and accurate guide to understanding organisational behaviour than the formal organisational structure.
- *Open systems* – organisations are systems of interdependent activities linking shifting coalitions of participants; the systems are embedded in (i.e. dependent on) continuing exchanges with and constituted by the environments in which they operate.

From the three more expansive descriptions above, it is possible to see how far systems design has evolved since the 1960s and 1970s. In keeping with this less rigid, scientific concept of organisations, McNamara (2003) on his extensive Minnesota Organisation Development Network (MDODN) website describes an organisation fairly informally as 'a group of people intentionally organized to accomplish an overall, common goal or set of goals'. He alludes to the fact that if an organisation were compared to a house, understanding and analysing how it is put together and operates would require the study of its dimensions. The former description represents closely the nature of construction organisations, which often form and operate in a transient environment and, to an extent, the latter broadly describes the need to study their resultant cultural components or dimensions (i.e. traits). Daft (2001: 16–21) develops this idea further, merging the 'architectural' concept with a socially based ideal, and places these dimensions into categories of contextual and structural, which he states describe specific organisational design traits, much in the same way that personality and physical traits describe people. His description of these dimensions is summarised below.

Contextual:

- Culture – the values and beliefs shared by all (of specific relevance to the next section in this chapter). McNamara (2003) also notes on his website that culture is often discerned by examining norms or observable behaviours in the workplace.
- Environment – the nature of external influences and activities in the political, technical, social and economic arenas.
- Goals – the unique overall priorities and desired end-states of the organisation.
- Size – the number of people and resources and their span in the organisation.
- Technology – the often unique activities needed to reach organisational goals, including nature of activities, specialisation, type of equipment/facilities needed.

Structural:

- Centralisation – the extent to which functions are dispersed in the organisation, either in terms of integration with other functions or geographically.
- Formalisation – the extent of policies and procedures in the organisation.
- Hierarchy – the extent and configuration of levels in the structure.
- Routinisation – the extent to which organisational processes are standardised.
- Specialisation – the extent to which activities are refined.
- Training – the extent of activities to equip an organisation's members with knowledge and skills to carry out their roles.

Daft (2001: 20) states that these 11 dimensions are interdependent and provide 'a basis for the measurement and analysis of characteristics that cannot be seen by the casual observer, and they reveal significant information about an organisation'. This latter view of organisations, together with McNamara's (2003) view of an organisation as a 'house' described earlier, are both relevant in the context of the construction industry and explain the basis of the qualitative research of selected construction organisations described later in the book that was undertaken to relate somewhat more generic organisational dimensions back to the largely empirical data obtained using the Denison Organizational Culture Survey (DOCS) and the Hong Kong Housing Department's Performance Assessment Scoring System

(PASS), which provides data on organisational effectiveness. The qualitative research is represented by the mini-case studies undertaken in Hong Kong and described in Chapter 9 of this book. The main purpose of these mini-case studies is to examine how the organisations whose culture is being measured are actually constituted, i.e. their cultural 'dimensions'. Having considered the various views on what an organisation 'is' and what it 'does' and based on the premise that this book, besides examining culture in construction organisations, also examines the major tenets of organisational culture and performance and the relationship between them, the simplified version of the definition of organisation from Daft (2001) is adopted throughout the book, which is that organisations are:

- social entities;
- goal directed;
- deliberately structured and coordinated activity systems;
- linked to the external environment.

Based on the contemporary view that successful organisations are operated on an open system approach and that 'older' management theories are in some way inadequate to cope with modern business environments and their problems and challenges, many authors advocate a 'contingency' approach as being the best type of management for the modern organisation (Daft, 2001: 24).

So to summarise this section, organisational theories have changed substantially over the past century; from scientific management (mechanical) to contingency theory (natural) as environments have altered from being more stable to more turbulent and this process has moved more rapidly over the past 20 years or so. Managers have had to restructure their companies to become 'learning organisations' in which communication and collaboration are supported by everyone becoming involved in identifying and solving problems. Adopting a contemporary systems view of construction organisations appears to lead to a position where no individual school of management thought can be seen as being completely right to tackle the ever-changing environment of construction projects, and this is particularly true for construction companies, where policies, processes and controls change often very significantly and those companies that are unable to adapt (i.e. 'learn' to cope and succeed in new environments) inevitably fail. Figure 2.2 (adapted from Daft, 2001) indicates the range of the shift that has occurred to organisational structures over the period described in the preceding section, and of particular interest are the changes from 'rigid' to 'adaptive' organisational cultures and also the concepts of the modern organisation being 'flat structured', collaborative, empowered and sharing information, traits which are closely connected with significant dimensions explored in Chapter 3 with regard to their significance to the construction industry in general and to Hong Kong construction companies in particular.

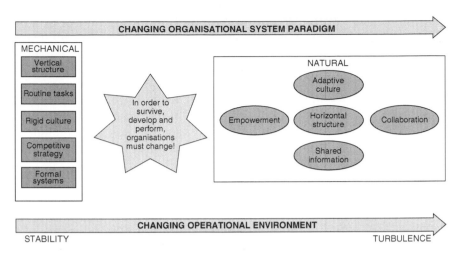

Figure 2.2 Historical shift in organisational structures (1930–2000).
Source: Daft (2001) adapted by Coffey.

What is organisational culture?

Culture – a historical view

The study of culture has its roots in the science of anthropology and these roots as they have developed have been impacted on by the areas of the social and human sciences. In his classic textbook *Cultural Anthropology: Tribes, States, and the Global System*, John H. Bodley (2000) writes that in 1872, the British Association for the Advancement of Science (BAAS) produced an inventory of cultural categories in the form of an anthropological field manual which listed 76 unordered culture topics. A more exhaustive manual, first published in 1938, was entitled *Outline of Cultural Materials* (usually abbreviated to OCM), which listed 79 major and 637 subdivisions, and is still in use today in its fifth edition (Murdock *et al.*, 1983). The BAAS was assisted by Sir Edward Burnett Tylor, an English anthropologist and Quaker, who was influential in the rise of interest in the study of anthropological science in England resulting from the popularity of a two-volume work of 1871, *Primitive Culture*, which documented his extensive researches on the mentality, customs and beliefs of primitive peoples. This book resulted in his election as a Fellow of the Royal Society and established him as the leading English anthropologist of his generation (Stocking, 1995). Considered to be of specific importance, his treatise on animism (the idea that all living things were produced by a spiritual force and have souls, and also that spirits and demons exist) made great advances in the understanding of primitive religions (Marett, 1936). Since the publication in 1881 of his subsequent book *Anthropology* (still considered to be relevant in terms of its cultural concepts and theories), enthusiastic debate has continued among anthropologists to evolve a universally acceptable definition of 'culture' and as to what are the most important technical attributes to stress in such a description (Mcbee and Warms, 2004) and the following subsection analyses the various viewpoints on the concept of culture and in particular organisational culture, which have been developed since Tylor's time.

Organisational culture

Given that the research described later in this book is based on the in-depth examination and assessment of the prevailing culture of a specific group of organisations, it is imperative to examine first what 'organisational culture' actually 'is', 'why' it has importance in an organisational context and then 'how' it can be investigated, with particular reference to the specific business context in which the book is set (i.e. the construction industry).

While some researchers have approached culture from the historical or behavioural viewpoints, others have adopted a more structural approach and some a symbolic perspective. Brown (1998: 9) attempts to classify these differences, at least in a rudimentary fashion, and posits that organisational culture is generally seen as a metaphor or as an objective entity. The latter is further subdivided into those who see such objective entities as a whole like Pacanowsky and O'Donnel-Trujillo (1982) and those who view it as a set of behavioural and cognitive characteristics such as Schein (1985) and Eldridge and Crombie (1974). The position taken in this book is that organisational culture will be viewed as an objective entity given that the organisations being scrutinised are all technological entities, staffed by scientists, technologists and tradesmen. Kroeber and Kluckhohn (1952) listed 160 definitions of culture and also tabulated the diversity of perspectives of culture, which had prevailed since Tylor's time and up until the 1950s. There appears to have been no subsequent attempt in the literature to rigorously update their original catalogue of 160 definitions. In addition, when examining the categorisations used by these authors, it is somewhat difficult to establish an all-embracing finite definition of culture that fits the discipline or theoretical perspective from which different researchers may wish to view the subject. In order to further illustrate the diversity and range of concepts and perspectives which have developed over the past 130 years and also to advance somewhat the current literature on the subject, Table 2.3 collects and summarises chronologically some of the most influential definitions that are extant in the culture and cultural anthropology literature, and this list also attempts to capture some previous (and current) collections of definitions which have been published on the internet by other researchers (Solot, 1986; Kunda, 1992; Garrity, 2003; Loubser, 2003; Maralgia *et al.*, 2003; Varenne, 2003). Following Table 2.3 is a short review on why the definitions appear to have subtly altered over time and a simple definition of culture is included, which has been used to inform the more specific definition of organisational culture which follows later in this chapter.

Table 2.3 Definitions of culture (1871–2003)

'is that complex whole which includes knowledge, belief, art, law, morals, customs, and any other capabilities and habits acquired by man as a member of society.' (Tylor, 1871)

'[is] that complex whole which includes all the habits acquired by man as a member of society.' (Benedict, 1934)

'means the whole complex of traditional behaviour which has been developed by the human race and is successfully learned by each generation. A culture is less precise. It can mean the forms of traditional behaviour which are characteristic of a given society, or of a group of societies, of a certain race, or of a certain period of time.' (Mead, 1937)

'embraces all the manifestations of social habits of a community.' (Boaz, 1940)

'obviously is the integral whole consisting of implements and consumer goods, of constitutional charters for the various social groupings, of human ideas and crafts, beliefs and customs.' (Malinowski, 1944)

'the culture of society is the way of life of its members; the collection of ideas and habits which they learn, share and transmit from generation to generation.' (Linton, 1945)

'[all the] historically created designs for living, explicit and implicit, rational, irrational, and non-rational, which exist at any given time as potential guides for the behavior of man.' (Kluckhohn and Kelly, 1945)

'the mass of learned and transmitted motor reactions, habits, techniques, ideas – and the behavior they induce.' (Kroeber, 1948)

'the man-made part of the environment.' (Herskovits, 1948)

'As a sociologist, the reality to which I regard the word "culture" as applying is the process of cultural tradition, the process by which in a given social group or social class, language, beliefs, ideas, aesthetic tastes, knowledge, skills and usages of many kinds are handed on ("tradition" means "handed on" from person to person and from one generation to another).' (Radcliffe-Brown,1949)

'[it] uses and transforms life to realise a synthesis of a higher order.' (Lévi-Strauss, 1949)

'The cultural category, or order, of phenomena is made up of events that are dependent upon a faculty peculiar to the human species, namely, the ability to use symbols. These events are the ideas, beliefs, languages, tools, utensils, customs, sentiments, and institutions that make up the civilization – or culture, to use the anthropological term – of any people regardless of time, place, or degree of development.' (White, 1949) 'consists of patterns, explicit and implicit, of and for behavior acquired and transmitted by symbols, constituting the distinctive achievements of human groups, including their embodiments in artifacts; the essential core of culture consists of traditional (i.e. historically derived and selected) ideas and especially their attached values; culture systems may, on the one hand, be considered as products of action, and on the other as conditioning elements of further action.' (Kroeber and Kluckhohn, 1952)

'an historically transmitted pattern of meanings embodied in symbols, a system of inherited conceptions expressed in symbolic form by means of which men communicate, perpetuate, and develop their knowledge about and attitudes towards life.' (Geertz, 1966)

'The sum total of the ways of living built up by a group of human beings, which is transmitted from one generation to another.' (*Random House Dictionary*, 1968)

'is all of those means whose forms are not under genetic control which serve to adjust individual and groups within their ecological communities.' (Binford, 1968)

'A society's culture consists of whatever it is that one has to know or believe in order to operate in a manner acceptable to its members.' (Goodenough, 1971)

'a system of symbols and meanings.' (Schneider, 1976)

'consists in the shared patterns of behaviour and associated meanings that people learn and participate in within groups to which they belong.' (Whitten and Hunter, 1987)

'the totality of the learned and shared patterns of belief and behavior of a human group.' (Aceves and King, 1978)

'learned behavior copied from one another.' (Steadman, 1982)

'the way we do things around here!' (Deal and Kennedy, 1982)

'means that total body of tradition borne by a society and transmitted from generation to generation. It thus refers to the norms, values and standards by which the people act, and it includes the ways distinctive in each society of ordering the world and making it intelligible. Culture is...a set of mechanisms for survival, but it provides us also with a definition of reality. It is the matrix into which we are born, it is the anvil upon which our persons and destinies are forged.' (Murphy, 1986)

'the patterned behavior and mental constructs that individuals learn, are taught, and share within the context of the groups to which they belong.' (Whitten and Hunter, 1987)

(*Continued*)

Table 2.3 Cont'd

'a set of shared ideals, values and standards of behavior; it is the common denominator that makes the actions of individuals intelligible to the group.' (Haviland, 1993)

'in its most basic form is an understanding of "the way we do things around here." Culture is the powerful yet ill-defined conceptual thinking within the organisation that expresses organisational values, ideals, attitudes, and beliefs. Culture is the strategic body of learned behaviors that give both meaning and reality to its participants.' (Cunningham and Gresso, 1994)

'consists of "learned systems of meaning, communicated by means of natural language and other symbol systems, having representational, directive, and affective functions, and capable of creating cultural entities and particular senses of reality." ' (D'Andrade, 1996)

'the learned patterns of behavior and thought characteristic of a societal group.' (Harris, 2004)

'We will restrict the term *culture* to an ideational system. Cultures in this sense comprise systems of shared ideas, systems of concepts and rules and meanings that underlie and are expressed in the ways that humans live. Culture, so defined, refers to what humans learn, not what they do and make.' (Kessing and Strathern, 1998)

'There is agreement that culture is learned from others while growing up in a particular society or group; is widely shared by the members of that society or group; and so profoundly affects the thoughts, actions, and feelings of people in that group that anthropologists frequently say that, "individuals are a product of their culture." ' (Bailey and Peoples, 1999)

'the set of learned behaviors, beliefs, attitudes, values, and ideals that are characteristic of a particular society or population.' (Ember and Ember, 2001)

'All aspects of human adaptation, including technology, traditions, language, and social roles. Culture is learned and transmitted from one generation to the next by nonbiological means.' (Jurmain *et al*., 2000)

'learned and shared patterns of thought and behaviour characteristic of a given population, plus the material objects produced and used by that population.' (Temple University, 2003)

Sources: O'Brien (2003), Smith (2003), Maraglia *et al.* (2003), Varenne (2003), University of Alabama (2003), Loubser (2003), Garrity (2003). Compiled by Coffey (2006).

A closer examination of the varied definitions in Table 2.3 at first glance appears to reveal that there has been very little shift from the thinking at the turn of the twentieth century, (i.e. culture is 'a complex whole' that incorporates all of the behaviours, habits and influences that an individual requires to fit into the social environment), to the more current contemporary perspective of culture as a set of values, traditions, beliefs and ideals that are learned from being a part of society. However, if we take a deeper look at how the view on culture has changed, we will begin to observe the subtle development of differing theoretical and academic approaches to the concept and thus also examine why this shift has occurred, linking it to the evolution of more general changes in organisational theory and management thinking over the same period of time. While the importance of systems and contingency theory in relation to describing organisations has already been discussed in an earlier section of this chapter, an examination of the following post-1950s schools of thought also reveals the effect that these theories had on the concept of the study of organisational culture:

- *Human relations (1950s to 1960s)* – influenced by Lewin, Herzberg and Maslow, and associated with Argyris, Bennis, Drucker and Likert; premised that organisations existed to serve human needs by motivating individuals and developing group dynamics which made organisations more efficient. The work done at this time in researching values, beliefs and attitudes had a great influence on the burgeoning interest in organisational culture.
- *Modern structural theory (1960s to 1970s)* – influenced by Foucault and Mintzberg and associated with Burns and Stalker, Lawrence and Lorsch, Katz and Kahn, and Harrison; considers organisations as rational, goal-oriented and mechanistic, and focuses on hierarchy and structure as epitomised in modern organisational charts. According to Brown (1998: 5), this school of thought has generally had 'rather minimal influence on the development of the culture perspective' other than the fact that many researchers see organisations as 'cultural systems' that are ripe for study, and more importantly it has separated theorists into two major groups, those who see culture as being something an organisation 'has' (i.e. as a variable) and those who see it as something an organisation 'is' (i.e. a root metaphor).
- *Power and politics (1970s to present)* – influenced by Pfeffer, Kantner and Mintzberg and associated with French and Raven, Porter and Handy; proposes that organisations are complex coalitions of groups and subcultures with competing values and preferences. This perspective aligns closely with the culture view that organisations are irrational entities that achieve their goals through a mixture of compromise negotiation, conflict dynamics and influence of leaders and subordinates.

The 'organisational culture' perspective is in effect itself a school of thought, which has developed out of the precedents mentioned above and is specifically associated with Schein, Trice and Beyer, Hofstede, Morgan, Alvesson, Martin and Smircich. It is predominantly interpretative and takes the form of either a rational or a critical model depending on the point of view of the researcher/author. The following operational definition of culture is used throughout the book:

> Culture is the system of shared beliefs, values, customs, behaviours, and artifacts that the members of society use to cope with their world and with one another, and that are transmitted from generation to generation through learning.
>
> (Bates and Plog, 1990: 7)

After examining the definitions contained in Table 2.3, it appears that this definition, which, according to Brooks (2002) is essentially in the anthropological tradition of Boas and his followers, encapsulates the basic descriptive elements, traits and commonalities of all other definitions and also provides a logical foundation for the definition of organisational culture which is evolved in a later section of this chapter, to differentiate its study from that of organisational climate.

The relationship between organisational culture and climate

Before moving on to review the literature covering the historical and current research concerned specifically with organisational culture and its perceived linkage to organisational effectiveness, an examination of the literature in relation to the studies of organisational 'climate' (since both organisational culture and climate are closely related) is at this stage of the chapter both relevant and necessary. The main reason for including this discussion here is to establish with a degree of confidence that what is being studied in this book actually is *culture* and not *climate*. In the early 1980s, when the new interest in cultural studies of organisations was surfacing, the difference between culture and climate was more defined; organisational climate has been defined as the:

> relatively enduring quality of the internal environment of an organisa-tion that a) is experienced by its members, b) influences their behavior, and c) can be described in terms of the values of a particular set of characteristics (or attitudes) of the organisation.
>
> (Tagiuri and Litwin, 1968: 27)

The climate is the 'ether' within which an organisation exists. Schwartz and Davis (1981: 32) stated 'one way to understand culture is to understand what it is not', their view being that it is not 'climate'. At that time, a prerequisite of most cultural studies required qualitative methodologies, while climate studies were usually situated in the quantitative paradigm (Denison, 1996). However, when examining the roots of both fields of study it became clear that they were quite different, culture research and writing emanating from the work of Cooley (1922) and Mead (1934) and climate studies from Lewin *et al.* (1939) and Lewin (1951). According to Brown (1998: 2), 'the real finding of the climate surveys was that a more sophisticated approach to understanding this aspect of organisation was required' and he opines that the organisational culture research movement filled this 'gap'. The extensive literatures on both topics, while highlighting distinct differences between the two areas of organisational study, also share some marked similarities, a phenomenon which has caused debate and a degree of confusion between researchers over the past 30 or 40 years (Denison, 1996). Schauber (2001: 4) describes organisational climate as being 'integral to and yet only a part of an organisation's culture, is easier to change than its culture ... found in the private language of the organisation, such as the conversations about work among staff during coffee breaks'. Figure 2.3 captures Schauber's (2002) view of the relationship between organisational culture and climate.

In 1996, Denison wrote an article that undertook to comprehensively examine and discuss the extant literatures of both topics and also 'their definition of the phenomena, their epistemology and methodology, and their theoretical foundations' (Denison, 1996: 619). He posits (ibid: 624) that 'culture refers to the deep structure of organisations, which is rooted in the values, beliefs and assumptions held by the organisational members' and contends that climate, contrastingly, 'portrays organisational environments as being rooted in the organisations value system'. Denison (1996) concludes that the two research traditions are not actually very different in terms of their phenomenological perspectives, but do have distinct differences in interpretation and also in the typical research methodologies applied to

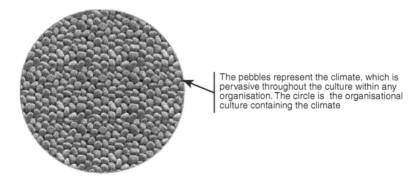

The pebbles represent the climate, which is pervasive throughout the culture within any organisation. The circle is the organisational culture containing the climate

Figure 2.3 The relationship between organisational culture and climate.
Source: Schauber (2002) adapted by Coffey.

each area. The author observes that there is a compelling argument for the use of both quantitative and qualitative methodologies traditionally aligned with each of the respective fields to be integrated when studying organisational culture, and that the literatures of both actually address a common phenomenon which is the development and impact of the social contextual perspective within organisations (ibid: 646). There were many other differences in the two study areas, which are summarised and located into their correct focus areas in Table 2.4.

Denison's findings are of particular significance in relation to the research methodology used in this book, since they support the mixed-method approach that is adopted to study the dependent variable (i.e. organisational culture). Table 2.5 also shows the Denison Consulting Group's analysis (1996) of the contrasting differences between the two literatures.

However, Wallace *et al.* (1999: 550) present a somewhat different view, namely that the literature has often overlooked the 'close and sometimes ambiguous relationship between organisational culture and climate' and cite Barker (1994), who concluded that 'there is evidence the two terms have frequently been used synonymously'. Their exploratory research conclusions found evidence to suggest that relationships between culture, climate and managerial values of organisations actually exist. This differed somewhat from the view of Litwin and his colleagues and contemporaries, who largely concentrated their research on attempting to establish and test the relationships that appeared to exist between organisational climate, managerial behaviours, norms and values (Litwin and Stringer, 1968; Tagiuri and Litwin, 1968). Some years later Schneider *et al.* (1994: 17) expanded Litwin's definition of climate and added to it their own definition of culture. These authors offered the following definitions: 'Climate – the "feeling in the air" one gets from walking around a company ... the atmosphere that employees perceive is created in their organisations by practices, procedures and rewards.' They define culture as 'the broader pattern of an organisation's mores, values and beliefs' and then integrate the two definitions as follows: 'Culture, then, stems from employees' interpretations of the assumptions, values and philosophies that produce the climates they experience' (1994: 17).

The section that follows will, first, revisit Denison's analysis and add to it some discussion of the more recent considerations of, and views on, the literature, including those of the authors cited by Wallace *et al.* (1999), and will then examine the most current perspectives on the paradigms represented by the two fields of study, before relating the extant literature to the research questions being examined in this book. When discussing organisational climate, it is important to distinguish from psychological climate which is also often studies in an organisational setting. Glisson and James (2002) define the latter as 'the individual employee's perception of the psychological impact of the work environment on his or her own well-being', whereas according to the same authors (2002: 769), 'When employees in

Table 2.4 Factors distinguishing the organisational culture and climate studies

Culture study focus (qualitative)	Main authors	Climate study focus (quantitative)	Main authors
Roots *Symbolic interactionism and socio-anthropology perspective*	Blumer, 1969 Cooley, 1922 Kluckhohn, 1951 Levi-Strauss, 1949 Mead, 1934	**Roots** *Lewinnian concept of individuals acting in context to produce behavioural patterns*	Campbell *et al.*, 1970 Evan, 1968 Forum Corporation, 1974 Forehand, 1968 Halpin and Croft, 1962 Lewin, 1951 Lewin *et al.*, 1939 Likert, 1961, 1967 Litwin and Stringer, 1968 Tagiuri and Litwin, 1968
Longitudinal studies of the evolution of social systems and socialisation	Martin *et al.*, 1985 Mirvis and Sales, 1990 Mohr, 1982 Pettigrew, 1979 Rohlen, 1974 Sathe, 1983 Schein, 1985, 1990 Siehl and Martin, 1981 Trice and Beyer, 1984 Van Maanen, 1979, 1975	*Studies on the impact of organisational systems on groups/individuals*	Ekvall, 1987 Joyce and Slocum, 1984 Koyes and DeCotis, 1991
Deeper level understanding of underlying assumptions	Dalton, 1959 Jacques, 1951 Kunda, 1992 Schein, 1985, 1990	*Organisational members' perceptions of internal practices and procedures*	Guion, 1973 James and Jones, 1974

(Continued)

Table 2.4 Cont'd

Culture study focus (qualitative)	Main authors	Climate study focus (quantitative)	Main authors
Individual meaning and symbolism and meaning	Geertz, 1971, 1973 Pondy *et al.*, 1983 Weick and Roberts, 1993	*Distinguishing climate from organisational facets (e.g. individual satisfaction/ organisational structure)*	Drexler, 1977 Guion, 1973 James, 1982 Johanneson, 1976 LaFollette and Sims, 1975 Lawler *et al.*, 1974 Payne *et al.*, 1976 Payne and Pugh, 1976 Schneider and Snyder, 1975
Inside views of organisations	Barley, 1983, 1986 Whyte, 1949	*Distinguishing 'psychological' climate from 'organisational' climate*	Glick, 1985, 1988 Guion, 1973 Hellriegel and Slocum, 1974 James and Jones, 1974 James *et al.*, 1988 Payne and Pugh, 1976
Analysis of cultural literature	Allaire and Firsirotu, 1984 Czarniawska-Jorges, 1992 Frost *et al.*, 1985, 1991 Geertz, 1973 Martin, 1992 Rosen, 1985, 1991 Smircich, 1983 Smircich and Calás, 1987 Trice and Beyer, 1992	*Examination of climate formation (i.e. 'where do organisational climates come from?' and climates as an 'outgrowth' of basic value systems)*	Ashforth, 1985 Poole, 1985 Poole and McPhee, 1983 Reichers, 1987 Schneider, 1987 Schneider and Reichers, 1983

Source: Authors/dates derived from Denison (1990, 1996) and tabulated by Coffey (2005).

Table 2.5 Perspectives of organisational culture and organisational climate research

Differences	Culture literature	Climate literature
Epistemology	Contextualised and ideographic	Comparative and nomothetic
Point of view	Emic (native point of view)	Etic (researcher's viewpoint)
Methodology	Qualitative field observation	Quantitative survey data
Level of analysis	Underlying values and assumptions	Surface-level manifestations
Temporal orientation	Historical evolution	A historical snapshot
Theoretical foundations	Social construction: critical theory	Lewinian field theory
Discipline	Sociology and anthropology	Psychology

Source: Denison (1996: 625).

a particular work unit agree on their perceptions of the impact of their work environment, their shared perceptions can be aggregated to describe their organisational climate.' Schneider *et al.* (1994: 17) express a similar view when describing employees' perceptions of climate, which they state are 'developed on a day-to-day basis ... *not* based on what management, the company newsletter, or the annual report proclaim – rather, the perceptions are based on executives' behaviour and the actions they reward'. Schneider (1990) edited a book which tried to draw distinctions between culture and climate; however, none of the writers of the individual chapters reached definite conclusions on the issue, In 1994, Schneider subsequently edited another series of studies, which this time critically examined the effects of climate and culture together and posited a view that if this dual research perspective is applied to the study of organisations, then the understanding of the behaviour of individuals within organisations can be improved or even advanced, and hence the organisation's overall performance in such areas as financial planning, marketing, human resource development and others will be strengthened.

These views have been expanded by the recent work of Sparrow (2001), who concluded that:

> both culture and climate assessments describe ways in which individuals make sense of their organisation and provide the context for organisational behaviour, allowing researchers, practitioners or consultants to

describe, explain or even predict why some behaviours are more effective than others for a particular organisation.

(Sparrow, 2001: 88)

However, a contrary view is expressed by Payne (2002: 166), who opines that there is no clear distinction between the constructs of culture and climate, and he extends this criticism to Schneider's earlier 1990 book. Denison (1996) in commenting on such arguments opines that rather than researchers engaging in a non-productive 'paradigm war', there is a case for climate-based research to inform culture-based research to give a clearer overall picture of a particular organisation's culture, which may then be related to assessable outcomes that result from possession of such a culture.

There has then, been, as may be seen from the brief examination of different organisational studies in this section of the literature review, a certain amount of controversy and dialogue with regard to the efficacy of using the culture or climate approach when investigating and measuring organisational dimensions, and it appears that the major difference between the assessment of organisational culture and organisational climate is that the former involves the analysis of typical organisational practices which produce measurable effects and the latter examines the views which participants hold about their organisations at a particular point in time. The research approach described later in this book is based on a synthesis of the available methodologies and has a framework that derives from both qualitative and quantitative methodologies in order to carry out an examination of organisational culture and its impact on organisational performance. In the next chapter, which reviews the specific literature on organisational culture, the various approaches to the assessment of organisational culture are examined in more depth and a synthesis of methodologies is proposed for rendering the most useful results, in the context of the area in which the author's own research is based.

Chapter summary

This chapter has reviewed and analysed in some depth the relevant literature that has covered the topics of culture and climate in organisations over the years and has prepared a foundation for the chapters that follow which will examine more closely the literature on the relationship between 'organisational culture' and 'effectiveness'.

3 Organisational culture studies

Chapter introduction

Having in the previous two chapters considered the part that organisational climate plays in the historical development of organisational research and what 'organisations' and 'organisational culture' *are*, as well as examining the importance and relevance of the latter to an organisation's persona and in relation to the study of organisations in general, it is now useful to determine the origins of organisational culture (as opposed to pure organisational) research and then to look at the reasons for the current interest in this area of study. Chapter 3 therefore investigates the development of organisational culture research, from its early beginnings in cultural anthropology, through its changing paradigms over the past 30 years or so and examines in detail the various ways of analysing organisational culture, ending with a description of instruments in use currently to do this.

The development of organisational culture research

Lewis (1996: 12) states that 'Organisational culture is but one dimension of the study of organisational behaviour and the concept is by no means a new one'. Lewis goes on to cite Potter (1989), who quotes Jaques (1951) as defining the culture (of a factory) as 'the customary and traditional way of thinking and doing things, which is shared to a greater or lesser degree by all its members, and which new members must learn, and at least partially accept, in order to be accepted into service in the firm' (Jaques (1951), quoted in Potter, 1989: 17). Jaques was one of the first researchers to use the term 'culture' with reference to the workplace. Now that the reader understands the concept of organisational culture as described previously, this chapter is the first of three that set out to further assist the reader in fully understanding the research that has been undertaken on organisational culture, and how data from various studies have been analysed and used as a management and change tool in various industries.

According to Brown (1998), current interest in organisational culture stems from four sources:

- organisational climate research;
- research into national cultures;
- human resource management research;
- failure of traditional approaches, with their emphasis directed at the rational and structural nature of organisations, to offer an explanation of organisational behaviour.

This interest among researchers evolved, as mentioned in Chapter 2, from the organisational climate studies of the late 1960s to mid-1970s and although, as stated earlier, there is a degree of linkage between the literatures of both, interest specifically on culture has a different emphasis and relates to different aspects of organisational behaviour than does its predecessor on climate. Before the mid-1960s, the intellectual/philosophical origins of the types of culture studies being carried out in the workplace owed much to the anthropological research into human societies and cultures which is known in its field-based approach as ethnology. Developing an earlier argument of Barley *et al.* (1988), Meyerson (1991: 256) notes that 'culture was the code word for the subjective side of organisational life ... its study represented an ontological rebellion against the dominant functionalist or "scientific" paradigm'. It is therefore worthwhile to consider the literature of this anthropological approach to cultural research, better known as 'cultural anthropology', since this is one of the major qualitative research methods used in examining organisational cultures.

Cultural anthropology

The *Penguin Dictionary of Psychology* (Reber, 1995: 42) defines 'cultural' anthropology as 'the sub-discipline within anthropology concerned primarily with the study of culture and the complex social structures which make up communities, societies and nations. The term itself is now used nearly synonymously with "social" anthropology'. According to Kottak (1994), the academic discipline of anthropology includes four main subdisciplines – socio-cultural, archaeological, biological and linguistic anthropology – and he refers to cultural anthropology as a synonym for 'socio-cultural anthropology'. He explains that cultural anthropology examines the cultural diversity of the present and the recent past and, together with sociology, shares an interest in social relations and behaviour. This description differs slightly from a more recent contemporary description (Ember and Ember, 2001: 5), which states that 'the three main branches of cultural anthropology are archaeology (*the study of past cultures, primarily through their material remains*), anthropological linguistics (*the anthropological study of languages*) and ethnology (*the study of existing and recent cultures*)'.

Although it is ethnology that appears to have the most relevance to organisational and business research, the cultural anthropology literature informs us that the subject has two distinct aspects: ethnography, (writing about people based on fieldwork) and ethnology (going beyond description to interpret the data collected based on cross-cultural comparison). According to Ember and Ember (2001: 8), ethnographers (*themselves a type of ethnologist*) usually spend time living among, talking with and closely observing the ethnic communities whose customs they are studying. This fieldwork then provides the data for an ethnography (a detailed description of the many aspects of observable attitudes, beliefs, behaviour and thought of the studied group). Kottak (1994: 34) notes that anthropologists normally enter the community they are studying and get to know the people, and states, 'I believe the ethnographic method and the emphasis on personal relationships in social research are valuable gifts that anthropology brings to the study of a complex society'. The counterpoint to this view is that when outsiders 'enter the community and get to know the people', there is always a danger that their intrusion into that cultural group may in fact alter its natural traits and therefore any subsequent analysis of that culture may accidentally carry bias not found in the natural existence of the group *without* the presence of the ethnographer. This is a crucial point to bear in mind in any consideration of the adequacy of culture research, as it touches on the important concepts of *emic* and *etic* approaches to studying culture. Pike (1967) first introduced the terms which may be summarised as follows:

- An emic account of human behaviour is delivered in terms meaningful (consciously or unconsciously) to the actor, i.e. the 'insider view'.
- An *etic* account describes behaviour in terms familiar to the observer, i.e. the 'outsider view'.

Kottak (1994: 14) acknowledges the importance of the emic/etic issue when he states 'In practice, most anthropologists combine emic and etic strategies in their fieldwork', and posits that in order to properly understand and interpret culture, ethnographers need to recognise the biases emanating from their own culture that may influence their research into other cultures. A detailed consideration of these aspects of organisational culture studies is described later, also explaining how the survey instrument used in the Hong Kong research study was adapted linguistically for use outside the culture in which it was developed. The pure ethnologist on the other hand is concerned with all aspects of culture in the contemporary world and attempts to present a perspective from which to understand modern society. Ethnologists stress the importance of observation together with the collection of actual data. In making comparisons between peoples and cultures, ethnologists need to differentiate between responses peculiar to the society being studied and those that are general to humankind. Ethnologists also study the dynamics of cultures (i.e. how cultures develop and change), together with the relationships between beliefs and practices within a culture (Ember and Ember, 2001), and if the data from a number of studies seem to suggest the existence of a general pattern between certain groups, this may be used as a basis for generalising about the relationship between different cultures, thus extending the knowledge and understanding of the way such cultures are organised (Bates and Plog, 1990). This latter view is also critical to this book, since one of the objectives of the research described later has been to determine what common cultural traits and strengths exist within an organisation that impact on the ability of that organisation to perform well within its relevant business sector. As will be discussed in the following section and developed in the methodology adopted for the specific research into the organisational culture of construction companies described later in this book, ethnography is an important component of modern organisational culture research studies.

The changing paradigms of organisational culture research

While the previous chapter examined the different definitions of culture and a general concept of organisational study, this section of Chapter 3 will examine the various directions that academic and business research has taken in relation to studying organisational culture over the past 20 years or so.

Meek (1988) opines that modern organisational culture studies grew out of the anthropological and sociological research fields, and this view is supported by Sackmann (1991) who states that 'Conceptions of culture in the organisational and management literature draw quite selectively from various anthropological and sociological sources'. Adopting a purely historical perspective of major research into organisational culture since the 1980s highlights the variety of approaches adopted in these studies as well as the range of opposing views which emanated from them. Denison (1996) talks of 'paradigm wars', and Martin and Frost (1996) refer to the 'organisational culture war games'; the latter authors also state that 'it is clear that any review of organisational culture research must respond to the existence of these disagreements' (p. 600).

So what were these various streams of research and where did the differences between them emanate from? Between the 1970s and 1980s, the interest in Japanese management successes and a growing dissatisfaction with traditional organisational analysis techniques had led some authors to criticise the mainstream quantitative-based, normalised scientific approach which prevailed in the United States and the United Kingdom. The main downside to this approach appeared to be that it was based on a rational model of human behaviour and was rooted in analysing numerical data drawn from highly structured research instruments that ignored the other fundamental faces of culture such as values, beliefs and shared attitudes.

Even the 'gurus' of early 1980s corporate research Peters and Waterman (1982: 4) state that 'if quantitative precision is demanded ... it is gained ... only by so reducing the scope of what is analyzed that most of the important problems remain external to the analysis'. The general feeling at the time that this alternative view to organisational research was developing was that the time had come to broaden the whole research playing field. Frost *et al.* (2000), referring to the opportunities presented by this new perspective, comment on the possibilities of doing more qualitative research and on being able to depart from technical and engineered approaches to studying organisations. In 1982, Peters and Waterman produced their popular book *In Search of Excellence: Lessons from America's Best-Run Companies*, and this became the first of a series of publications aimed at managers and MBA students to advocate the use of an appealing 'strong' (i.e. 'unified') corporate culture approach to improve organisational effectiveness. Other authors who produced similar books supporting the efficacy of the strong culture/effective cultural leader ideals at this time were Ouichi (1981), Pascale and Athos (1981), and Deal and Kennedy (1982), and for a while culture became a much sought-after product in the management sector and fuelled a whole new area of operations for business consultants. The development of culture–performance link studies is specifically addressed in Chapter 4. This whole movement of research interest was termed by some as the 'value engineering' approach (Martin and Frost, 1996: 602), and it was not long before it crossed over from the consultancy to the academic research field. This led to several studies that largely defined culture as an internal and consistent phenomenon and which drove organisation-wide consensus based on common shared values and beliefs (Pfeffer, 1981; Pondy *et al.*, 1983; Sergiovanni and Corbally, 1984; Sathe, 1985; Pennings and Gresov, 1986; Enz, 1988; Ott, 1989). In this academic research stream the approach was described as 'integration' research (Martin, 1992), and the historical context of this approach is described by several authors in terms of the benefits of the strong and unified culture model in improving financial bottom lines, a view which will be examined in Chapter 4 when the topic of measuring organisational effectiveness is considered.

As this integrated approach was emerging, a contrasting approach was being taken by other researchers, which would become known as the differentiation perspective (Martin and Meyerson, 1988). Foremost among such research studies were those produced by Van Maanen and Barley (1984) and Van Maanen and Kunda (1989), which noted the misalignment of stated attitudes and actual behaviour (i.e. the inconsistency between the managerial 'talk' and the operational 'walk') in organisations, and most importantly the differences of interpretations of events between one person and another. The differentiation movement challenged the hitherto supported research findings that organisations possessed cultures that were composed of clearly seen values, shared perceptions of organisational vision and enactment of organisational objectives in a consolidated and integrated manner.

Their view was one of highly complex organisational ethnographies, full of conflict and inconsistencies; thus they held the antithesis of the integration perspective. Even within this differentiation perspective, there were further subdivisions of which the two main streams were those which appeared to ignore pluralism within culture (i.e. offering a single interpretation of what is observed without challenging managerial theories) found in the research of Barley *et al.* (1988), Martin and Siehl (1983) and Louis (1985), and the other based on a more critical (i.e. anti-management approach) perspective (Rosen, 1985; Van Maanen, 1991; Alvesson and Berg, 1992; Alvesson, 1993). Such differences of approach within the same research category sometimes led to confusion among the major proponents, so that there were examples of the differentiation researchers criticising studies which highlighted the differentials of organisations (Alvesson, 1993). During the 1980s, according to Martin and Frost (1996: 605), 'literally hundreds of integration studies were published ... differentiation scholarship had been outflanked by a value engineering perspective'. Barley *et al.* (1988) noted that most studies, including those emanating from within academia, had primarily taken a managerial emphasis, and Calás and Smircich (1987) proclaimed that the interest in culture as a focus of organisational study, although 'dominant', was also 'dead'. The new 'paradigm wars' between integration/differentiation methodologies eventually evolved into a different sort of battle as researchers began to concentrate their discussions and differences into the area of competing methodologies, often with 'firmly held [but different underlying] epistemological beliefs' (Burrell and Morgan, 1992) for researching into organisations. This battle was particularly prevalent in the US where the major research model was managerially based, quantitative and integrated. Qualitative researchers responded to this situation by returning to fundamental ethnographic methods (i.e. longitudinal, full immersion and participative enquiry within cultures) and attacked short-term, interview-based qualitative studies, branding them as 'smash and grab' ethnographies (Sutton, 1994). This period is a critical one in relation to this book, as it was the time when the choice of paradigms and methods became crucial to particular researchers getting their studies published, a primary consideration when viewing how the research develops and moves forward. In an attempt to overcome the criticisms now being levelled at some of the qualitative studies, some researchers developed quantitative measures of culture, often drawing on surveys of the type previously used in climate studies. In the forefront of this neo-integrative movement over the 10-year period were Denison (1990), together with Gordon (1985) and Ouchi and Johnson (1978), who all claimed to have established a significant relationship between culture and financial performance, an area which will be discussed in more detail later in this book. This debate on the respective efficacy of using quantitative or qualitative methodologies to study culture has continued (Hassard and Pym, 1990; Schneider, 1990; Denison, 1996) and seems unlikely to be easily resolved

(Martin and Frost, 1996). Towards the end of the 1980s a new perspective emerged, named as the fragmentation approach, which 'conceptualizes the relationship among cultural manifestations as neither clearly consistent nor clearly inconsistent ... in the fragmentation view, consensus is transient and issue specific' (Martin, 2002: 94). Fragmentation studies of organisations thrived on a perspective which made no attempt to draw conclusions on how to control or normalise the 'cultural jungle', but rather highlighted the contradictions and resultant confusion in organisations that gave rise to ineffectual cultures and impacted adversely on performance and business success. Martin (2002) describes the state to which all of these developments had brought the study of organisational culture by the end of the 1990s as follows:

> Thus, integration, differentiation, and fragmentation researchers defined culture in a particular way and then designed studies which made it more likely they would find what they were looking for. This problem of tautology explained, to a large extent, why three traditions of research on ostensibly the same topic [organisational culture] could produce such conflicting empirical records.
>
> (Martin, 2002: 609)

Frost *et al.* (1991) and Martin (2002) have advocated an approach which combines all of these apparently conflicting research perspectives, termed the three-perspective framework, arguing that any in-depth organisational research will generate results which may be viewed (depending on the particular perspective of the researcher) as representing any (and thus all) of the three individual theoretical perspectives.

The most recent theoretical approach to the impact on studies of organisational culture comes from the postmodernist school, and while this perspective is outside the scope of the research stream described later in this book, a short description follows in order to complete this section examining the development of organisational culture research since the 1980s. According to Rowlinson and Procter (1999: 378),

> As with the concept of culture, post-modernism is difficult to define, and critics complain about the lack of definition. It is probably best thought of as an umbrella term for various discussions concerning 'language, culture, the subject, rationality, and writing'

Post-modernism challenges the very roots of modern scientific thinking and research as it draws attention to disorder and offers contradictory complex viewpoints which present reality as a 'series of fictions and illusions' (Martin and Frost, 1996: 611). Postmodernist researchers rely upon the 'use of disciplines such as linguistics, psychoanalysis, anthropology, literary criticism, and history' (Turner, 1990: 87). The research outcome of the postmodernist deconstructive approach is often a critique that attempts to 'overturn a disciplinary and prejudicial order through the articulation of ambiguity and contradiction from the margin ... but does not seek to coordinate this difference into a fresh hierarchy' (Jeffcutt, 2004: 18–19).

Discussions and arguments on the best methodologies to use to investigate, and perspectives from which to view, organisational culture continue, and in order to try to clear the air of such contrary views, many authors now take the position of Martin and Frost (1996) who advise,

> Let's find a way or ways to focus as much energy as we can to keep the study of culture free of destructive conflict, so that we can collectively imbue it with characteristics that will invite us to do our best work.
>
> (Martin and Frost, 1996: 616)

Analysing organisational culture

Organisational culture studies – a review

Considering the growth of interest by managers in the study of organisational culture during the 1980s, it was not surprising that by the early 1990s researchers had begun to review the body and state of knowledge that then existed in the extant literature and this produced a spate of criticism of what had gone before. Hofstede *et al.* (1990) when describing earlier studies observed that 'the literature on organisational cultures consists of a remarkable collection of pep talks, war stories, and some insightful in-depth case studies'. Thompson and McHugh (1995) opined that much of the earlier research relied heavily on the perspectives of 'leaders' and that the persistent criticisms of 'lack of rigour in research methodology' were justified. Denison (1990: 2) shared this view that 'none of these studies, for example, have followed a pattern of first formulating a set of criteria that would capture their "theory" of what makes organisations tick, and then going out to see whether they were right or wrong'. He further posited that there seemed to be evidence of another potential distortion in studies of the 1980s due to the focus of previous research on 'exemplary' organisations. He states (1990: 3) that 'none of these authors studied firms that failed or did poorly ... presumably, if a "theory" is correct, the factors that contribute to success should also serve to separate successful and unsuccessful firms'. Hollway (1991) concluded that there was a 'one-sidedness' of views of the fundamental issues surrounding the study of organisational culture which emanated from the 'management' perspective taken in texts of the 1980s and also 'the lack of theory' informing the approach to most research up until that time. His view was shared by others, notably Sackmann (1991), who opined that 'a critical review of the literature reveals ... that empirically based knowledge about culture in organisational settings is rather scarce and spotty', and this opinion was echoed by Willmott (1993), who states that 'considering the volume and influence of books and articles that celebrate corporate culturism in its various guises, there is a remarkable dearth in serious, critical analysis of this phenomenon'. Denison (1990: 30) argues that one of the major reasons that organisational culture studies are so open to such critical attack is due to 'the lack of evidence that scholars and researchers have presented in defense of their positions ... this problem appears even more serious when it becomes apparent that academic researchers have largely neglected the functional aspects of organisational culture'. Gray (1998) writes that based on such examination, a degree of scepticism is justified, and summarises the concerns of the critics by noting that 'If research is largely "sponsored" by one stakeholder group, if the evidence is heavily anecdotal, and if the subject organisations tend to be successful rather than the unsuccessful ones, what is learned may be inaccurate or misleading'. It appears then, judging by these critical comments regarding the failure of culture studies (the majority cited having been conducted since the early 1980s), to empirically 'deliver' robust conclusions that can be generalised (i.e. only the 'excellent' companies are examined) that the research theories, methodologies, samples and study designs were all in need of careful review and a degree of reframing.

Just as criticism was emerging by the end of the 1980s on both the concept and basis of organisational culture research, so a discussion was also building concerning methodologies used for studying the cultures of organisations. The more generic debate on competing research paradigms, as well as the justification for the use of a mixed methodology approach adopted for the research study described in this book, are reviewed in Chapter 6, but specific discussion on the use of qualitative or quantitative methods for studying culture and how this played a part in developing techniques and tools used to evaluate its strength and efficacy within organisations is also pertinent to this chapter. Questions concerning the methods for measuring culture had already been raised by discussions in studies carried out in the mid-1980s (Schein, 1984; Sashkin and Fullmer, 1985), and the need for more investigation into the feasibility and direction of cultural change had been highlighted by articles in prominent business magazines (Uttal, 1983; *Business Week*, 1984). The following section examines organisational culture studies as a specific area of research, i.e. *not* within the context of organisational culture–performance studies, which are discussed in depth in Chapter 5. An overview of some of the existing approaches to the study of culture in organisations may be found in an article by Sackmann (1991), which presents a methodology designed to overcome some of the deficiencies of previous studies. This author also discusses the advantages/disadvantages of the various methodologies used to study culture, ranging from mail-out survey questionnaires to participant observation reports, and proposes the use of an inductive methodology that serves as an appropriate compromise located somewhere between a detailed ethnographical and a quantitative questionnaire approach.

The advantages of combining methodologies when studying organisational processes and attributes are supported by Cooke and Rousseau (1988), who view them as 'complementary' and both capable of 'producing cumulative bodies of information for assessment and theory testing'. However, the authors stress that quantitative approaches 'may be more practical for analyzing data-based change in organizations', data-based change being an attribute which they opine is a powerful strategy contributing to organisational development (Cooke and Rousseau, 1988: 246). As highlighted in the previous chapter, there have been a considerable number of cultural studies that have produced a range of underlying definitions of organisational culture and so the measurement of the construct of organisational culture has been a major challenge to researchers operating in the quantitative paradigm. This has been further exacerbated by the lack of existence of any generally accepted, universal or comprehensive framework with which to measure and compare such organisational cultures. Research studies that were informed by a 'qualitative orientation' with 'verification' definitely set in the quantitative paradigm and which in my view have changed the direction of culture-based research in organisations are the 1983, 1988 and 1990 studies of Geert Hofstede and colleagues. Hofstede *et al.* (1990: 287), using the results of these studies, lead the reader through three fundamental questions and then provide answers to them as follows:

1   Can organisational cultures be 'measured' quantitatively, on the basis of answers of organisational members to written questions?
    Answer – Yes, as long as the differences between one organisation and another are due to the variance in members' views.
2   Which independent dimensions can be operationalised and used for measuring culture?
    Answer – Any discrete dimensions which had been gathered from analysis and were covered in the literature (unlikely that there were any undiscovered dimensions).
3   To what extent can measurable differences between organisations be due to unique features of each organisation?
    Answer – As organisational cultures are partly determined by nationality, industry and task and also related to structure and control systems, sufficient unexplained variance should exist between these measures to support the specific uniqueness of individual organisations.

Of the outcomes, Hofstede *et al.* (1990: 314) state: 'Our results, we believe, contribute to a demystification of the organisational culture construct, changing it from a passing fad into a regular element of the theory and practice of the management of organizations.' Wilderom *et al.* (2000: 201–202) do not refer to any of Hofstede's studies but do question whether *any* of the major studies they reviewed actually measured 'organisational culture', and identified the areas of their critical focus (i.e. the dimensions measured). Hawkins (1997: 436–438), in a seminal paper fostering a breath of fresh air in this paradigmatic debate, presents a new (postmodernist) vision of the shape of organisational culture research for the future. Hawkins (1997: 437), referring to Pettigrew (1990) as summarised by Mohan (1993), exposes the major challenges in this field as being due to

> the fact that culture is a multilevel construct that may be approached at a highly abstract, metamorphic layer or a more tangible artifactual level ... most investigations are capable of only a partial analysis that captures cultural snapshots of a distinct workplace group.

Hawkins believes that both researchers and consultants need to draw on a much wider range of academic disciplines and work in closer collaboration with members of organisations to discover the cultures (he states that participant observation cannot lead to full discovery). He rejects the 'neat meta-narrative' perspective of cultures, which he posits are a complex 'interlocking tapestry', and urges consultants to draw upon a richer menu of cultural change interventions to suit this tapestry. Hawkins (1997: 438) concludes by stating: 'Organisational culture needs to further develop itself as an action science that bridges the academic world of reflection and understanding and the consultancy world of enabling complex change and development. There is still much to be done'.

Survey instruments

Cooke and Rousseau (1988) utilised a self-scoring survey instrument, the Organisational Culture Inventory (OCI), which had been developed several years earlier (Cooke and Lafferty, 1983, 1986), to test the hypothetical norms and expectations of members of organisations and ascertain by measurement that these were being used in the same way throughout the organisation. The OCI was complementary to the Life Styles Inventory (Lafferty, 1973) which measured self-perceptions of individuals in relation to 12 causal thinking styles that were believed to have an impact on managerial effectiveness, the quality of interpersonal relations and individual well-being (Cooke and Rousseau, 1983). Table 3.1 gives the key to the OCI circumplex and directly relates the management styles operated in organisations to the concerns of employees, together with their orientations (i.e. interactions/interrelationship).

Using these dimensions, the OCI 'provides a profile of an organization's operating culture in terms of the behaviours that members believe are required to "fit in and meet expectations" within their organization'. The outcomes of the 1988 study indicated that behavioural norms vary across organisations and do so in a way related to differing management styles and that such norms are 'amenable to quantitative assessment' of results obtained

Table 3.1 Key to the OCI circumplex model

Style	Concerns	Orientations
Constructive	• Achievement • Self-actualisation • Humanistic/encouragement • Affiliation	Outside the three 'styles' are four orientations which merge at no particular fixed points reflecting the interrelationship of styles and interaction of the concerns. These are: • satisfying needs; • people orientation; • security needs; • task orientation.
Passive/defensive	• Approval • Convention • Dependence • Avoidance	
Aggressive/defensive	• Opposition • Power • Competition • Perfection	

Source: Human Synergistics International (2004) adapted by Coffey.

from using a qualitative model such as the OCI (Cooke and Rousseau, 1988). The OCI 'circumplex' originally used in the Cooke and Rousseau (1988) study has changed little today and is configured of 12 distinct and interrelated task-based/interpersonal substyles, their placement being related to their anticipated degree of association, and these fit within three perceived main styles (constructive, passive/defensive and aggressive/defensive). The OCI was revalidated in a further study (Cooke and Szumal, 1993) and has been used extensively for academic research and management consultancy studies since it was first devised. Cooke and Szumal (1993) have stated that the discriminant validity of the OCI may be flawed and they posit that while this may be the result of a weakness in the design of the OCI, it could also be due to the fact that the 'norms' for the general culture styles are somewhat loosely linked in certain settings. Recently, the model has been operated using new analyses of OCI data, and forms the basis for a proposed model of 'how culture works' (Cooke and Szumal, 2000). The findings and results from this new use of the model were as follows:

- Behavioural norms measured by the OCI are related to individual, group and system-level criteria of effectiveness.
- Such norms are also related to antecedent variables (which may be used to lever change).
- The theoretical model derived from this research study explains why the operating cultures of organisations are often inconsistent with the mission, as well as with the espoused values of the members.
- The latter finding is also connected to the phenomenon of the relationship of culture and effectiveness not always being as expected.

According to the home webpage of *Human Synergistics International* which now markets this tool, it measures 'how things are done around here ... OCI also measures certain key outcomes – individual member satisfaction, commitment, role clarity, role conflict and perceptions of the organization's service quality' (Human Synergistics International, 2004: 1). The same webpage defines culture thus:

> Culture refers to the set of particular behaviours (called 'behavioural norms') that an organization's members believe are expected of them if they are to 'fit in' and 'survive'. These behavioural norms lead to patterns of behaviours and attitudes that guide the way members approach their work and interact. These patterns can be positive and productive ... or not.

Ashkanasy *et al.* (2000) review the development of questionnaire-based surveys used to evaluate culture and conclude that such instruments have a major part to play in culture research, especially when combined with, or supported by, other qualitative data sources, a conclusion which supports earlier work by Reichers and Schneider (1990) and Rousseau (1990). Ashkanasy *et al.* (2000) use Schein's (1985) model of culture as a classification framework for reviewing specific questionnaire measures. His model of culture is shown in Figure 3.1 and of the three levels shown in this model, 'basic underlying assumptions' represents the deepest level of culture, while 'espoused values' and 'artefacts' represent incrementally shallower levels.

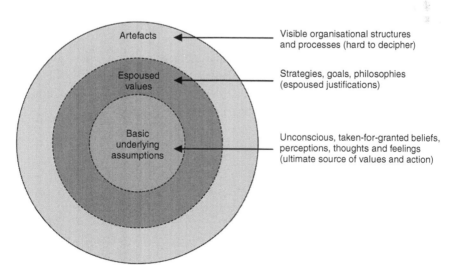

Figure 3.1 Schein's Model of Organisational Culture.
Source: Schein (1992) adapted by Coffey.

Drawing on Smircich's (1983) conclusion that culture is either something an organisation 'is' (deeper level) or 'has' (shallower level) and confirming that as the focus of their review is on survey instruments, they are thus limiting their consideration to the latter, Ashkanasy *et al.* (2000) accept that 'the shallower layers of culture are more explicit and can be appropriately studied using a structured and quantitative approach', echoing the views of Ott (1989) and Rousseau (1990). Appendix A, adapted from an earlier paper (Broadfoot and Ashkanasy, 1994), conveniently arrays the various organisational culture instruments used by other researchers between 1980 and 1992, which facilitated the development of these authors' own Organisational Culture Profile instrument in 1994. Ashkanasy *et al.* (2000) determined that culture surveys could be classified as either 'typing' or 'profile' scales with three subcategories of profile scales, and these were shown to be operable on the two outer layers of the culture model described earlier (Schein, 1985), i.e. artefacts and espoused values. 'Typing surveys' use instruments that classify organisations into cultural 'types'; for example, using the metaphors of Handy (1993) and Harrison (1972), the four prevalent types of culture and the organisational structures that best support them are shown in Table 3.2.

Handy (1993: 181–191) states importantly that the choice of an appropriate structure to support high levels of performance will be determined by:

- history and ownership;
- size;
- technology;
- goals and objectives;
- the environment;
- the people.

Handy (1993: 209) also observes that within every organisation, the (business) activities can be divided approximately into four sets:

- steady state;
- innovation;
- breakdown;
- policy.

Profiling surveys describe organisations based on a measurement of strengths and weaknesses as revealed by obtaining responses to individual organisational members' views on beliefs and values. The various cultural dimensions generated from these enquiries are then used to 'profile' the culture of the organisation(s) being examined. According to Ashkanasy *et al.* (2000: 135), 'Profiling surveys can be further divided into three subcategories: effectiveness surveys, descriptive surveys, and fit profiles.' The authors state that effectiveness surveys are the most prevalent in the field,

Table 3.2 Types of culture and supporting
structures

Cultures	Structures
Power	Web
Role	Temple
Task	Net
Person	Cluster

Source: Handy (1993), Harrison (1972), adapted
by Coffey.

as they measure the levels of values that appear most connected with high
levels of organisational effectiveness. Thus, the concept of a profiling survey
which measures effectiveness is highly relevant to this book as the DOCS,
described in detail later in Chapter 5, is such a profiling type instrument.

Chapter summary

This chapter has given the reader an overview of the extant literature on the
development of organisational culture surveys and it has also introduced the
parallel literature on assessing organisational climate. The chapter attempts
synergy between both concepts within the methodology of the construction
sector-related research study described later in this book.

4 Measuring organisational performance and effectiveness

Chapter introduction

In terms of tracing their historical development, the literatures related to the measurement of performance and the determinants of organisational effectiveness are closely intertwined. This chapter now examines the various theories and measures of business performance and organisational effectiveness, and attempts some comparison between them.

Organisational performance

According to Neely *et al.* (1995), numerous frameworks have been used by organisations since the beginning of the twentieth century to ascertain and measure their corporate performance. Until the mid-1980s the standard methodology revolved around ascertaining the financial strength, internally or through external third-party evaluations (e.g. Fortune 500 or 1000, America's Best 100 Companies, Goldsea 100), performance or success of organisations as the only measure of business performance. In the early 1900s, DuPont was using a pyramid of financial ratios, each of which was linked at different organisational levels to return on investment. Organisations continued to use such measures well into the 1970s and 1980s, although General Electric did implement a more balanced set of performance measures (i.e. not purely financially based) during the 1950s (Bruns, 1998). However, it was not until Johnson and Kaplan (1987) produced their comprehensive review of the evolution of management accounting systems (which revealed the failure of financial reporting instruments to reflect changes in the strategic focus and competitive environment within modern organisations) that the performance measurement focus began to change. Interest in developing alternative performance measures continued to grow during the late 1980s and early 1990s, and Keegan *et al.* (1989) proposed a simple balanced performance measurement matrix that emphasised the need to measure the 'cost', 'non-cost', 'internal' and 'external' dimensions of organisations. Lynch and Cross (1991) describe the Strategic Measurement and Reporting Technique (SMART) pyramid developed by Wang Laboratories that cascades its measures down to reflect both corporate vision and internal and external focus. Fitzgerald *et al.* (1991) classified measures into two categories as follows:

1   Results-focused: competitiveness and financial performance.
2   Results-determinants: quality, flexibility, resource utilisation and innovation.

These findings were important, as they indicated that in order to achieve desired success outcomes, organisations were first required to identify effective drivers of performance. Brown (1996) developed his Macro Process Model which demonstrated how organisational inputs incrementally affected not only the performance of organisational processes, but also impacted on the goals and objectives at the top levels of organisations (Kennerley and Neely, 2002). Probably the most popular and well used of the performance measurement frameworks has been the Balanced Scorecard developed by Kaplan and Norton (1992, 1996), and this is described in detail in the next section. Kennerley and Neely (2002) in their brief review of the performance measurement literature and contemporary models attempt to summarise and synthesise the main components of an effective performance measurement system as follows:

• must provide a 'balanced' picture of the business;
• needs to present a succinct overview of the organisation's performance;
• should be multi-dimensional;
• requires comprehensiveness;
• must be integrated both across the organisation's functions and through its hierarchy;
• business results need to be seen as a function of the measured determinates.

Kennerley and Neely (2001) develop their own integrated performance measurement model, the Performance Prism, a stakeholder-focused instrument designed to satisfy the parameters described above. The relevance of this perspective of performance consideration will be discussed in Chapter 5 in relation to the HKHD's Performance Assessment Scoring System (PASS) in terms of its effectiveness as a performance measurement tool. Up until this point, the section has briefly examined the history of development of performance measurement systems and provides a foundation for the next section, which examines in depth the operational concept of organisational effectiveness.

Organisational effectiveness

A general business-based definition is given by management consultants Mercury Group Inc. (2002) as follows: 'organisational effectiveness is achieving optimal results at the lowest possible cost. There are many factors that drive organisational effectiveness, and each impacts the others – processes, culture, measurement and reward, staffing, etc.' This is interesting in terms of the aspects of effectiveness it identifies as one fairly contentious issue in relating organisational culture and effectiveness, already mentioned in the preceding section, to the choice of what measures of an organisation's effectiveness are used and how these are then compared in order to establish what reflects best or worst performance. Opposing views within the organisational theory field, which represent this discussion, are those of Steers (1975) and Zammuto (1982), who have opined that the methodologies and instruments used to measure effectiveness constitute a major problematic issue. Hitt (1988), on the other hand, supports the measurement of organisational effectiveness as the major constituent of an organisation's change and improvement process. As discussed in the preceding section of this chapter, much of the previous research on performance measurement, and hence organisational effectiveness, undertaken since the early 1970s has relied on the use of financial measures to determine performance levels; typically, individual accounting indices such as return on investment (ROI), return on assets (ROA) and return on equity (ROE) have been used (Thune and House, 1970; Grinyer and Norburn, 1975; Karger and Malik, 1975). Longitudinally measured multiple financial accounting indices (to control for effects of inflation and changing industrial environments) were used in some later studies (Rumelt, 1974; Hill *et al.*, 1982), and the capital asset pricing model performance indicators evolved by Lubatkin (1983) were also used as a basis for research by Hitt and Ireland (1984). Despite the popularity of such models and indicators as well as the fact that most of these individual measures were highly inter-correlated, according to Hitt (1988) the use of financial indicators to represent organisational effectiveness has been criticised over the past 20 years. Hitt (1988: 30) states that

> they are subject to distortion of actual of actual company performance in any given time period. Discretionary depreciation methods and unusual charges or windfalls may distort performance. The number of stock shares outstanding affects earnings per share. Additionally these measures are rarely adjusted for risk.

Turning to the capital asset pricing model, some authors have suggested that this theoretical corporate performance indicator provides a better predictor and measure of long-term organisational effectiveness, as it incorporates risk adjustments and gauges effectiveness potential based on analysis of market returns to shareholders from the minimal risk asset gathering, thus addressing the deficiencies found in the use of pure financial indicators (Langetieg, 1978; Reilly and Brown, 1979; Weston and Brigham, 1982). However, according to Bowman (1980), this measure too has been criticised for assuming that the relationship between risk and return is always a positive one. These financial 'goals' were by this time considered to be unsuitable due to increasing evidence that organisational effectiveness could be adversely affected if inadequate goals were selected (Miles and Ezzell, 1980; Mohr, 1983). Summarising the apparent ineffectiveness of such financial-based measures of organisational performance, Hitt (1988: 30) notes that they 'have deficiencies as "true" indicators' and their use 'does not capture the essence of organisational effectiveness'.

Although much of the management interest was focused on using the measures described above to gauge how well (or badly) their organisation and others were performing, an alternative approach which was also developing at this time was the use of goal-based attainment measures, seen through the perceptions of managers (Paine and Anderson, 1977). Kirchoff (1977) recommended that policy research needed to move into more complex measurements of multi-goal achievement and this move away from sole reliance on financial measures was supported by Bourgeois (1980), who suggested that agreement on the strategies needed to achieve goals was an indicator of 'healthy' corporations and Peters and Waterman (1982), who stated that companies with broader managerial goals performed better than those with purely financial goals. Smith (1998) notes that eventually researchers became disenchanted with difficulties in using goal achievement model measures and investigated and developed several alternative models. Systems-based models focusing on a broader set of variables ultimately replaced goal-based benchmarks, but according to Mohr (1983) were often incorrectly developed setting the wrong targets, thus causing quite negative results to organisational performance. However, certain authors such as Smith (1998) still maintain that the traditional 'goal achievement' model remains the most popular, and he posits, 'any evaluation of organisational effectiveness must include an evaluation of relevant performance compared to that of competitors' (Smith, 1998: 34). More will be discussed on this statement in relation to how to measure the effectiveness of construction company performance in a later section.

Research studies in organisational effectiveness

Various effectiveness measures have been used in organisational research studies and these include:

- minimisation of regret (Keeley, 1978);
- winning percentages of ball teams (Eitzen and Yetman, 1972; Allen *et al.*, 1979);
- total production lost due to errors, low morale and anxiety (Pennings, 1976);
- employee error rates (Petty and Bruning, 1980);
- fire department outcome variables (e.g. incidence of fires, dollar losses due to fires and total fire loss per capita) (Coulter, 1979).

The literature reveals that most of the studies carried out between the mid-1960s and the early 1980s, as discussed above continued to use the more easily accessible rational goal performance measures such as:

- ROI (Hirsch, 1975);
- profit indices, together with production and quality indices (Kimberly and Nielsen, 1975);
- ROA (Snow and Hrebiniak, 1980);
- profit, profitability and stock prices (Weiner and Mahoney, 1981).

In the whole field of organisational research being undertaken at this time, there appeared to be very little consistency being applied across such studies; neither was there any consensus as to what measure (or combination of measures) was the most appropriate for ascertaining organisational effectiveness. Despite this divergence of measurement approaches in the published studies, the literature indicates that several alternative approaches were actually being recommended (Steers, 1975, 1976; Keeley, 1978; Connolly *et al.*, 1980; Cameron, 1980, 1981). As had been the case with the lack of consensus among both researchers and managers on defining organisational effectiveness, a similar dilemma existed in relation to agreement on the most effective measurements of the phenomena (Steers, 1975; Molnar and Rogers, 1976; Cunningham, 1977; Hrebiniak, 1978; Bourgeois, 1980; Tsui, 1981). Goodman and Pennings (1977) and Cameron (1984) identified that there were five major models that appeared to be appropriate to measuring organisational effectiveness, and these are shown in Table 4.1.

Some researchers questioned whether, in view of the lack of agreement on such fundamental questions as definitions and measures, organisational effectiveness should even be researched at all (Hannan and Freeman, 1977; Bluedorn, 1980). Emanating from such concerns over appropriateness of measures was what Rojas (2000: 97) describes as 'the multiple-constituency models designed to measure effectiveness not only internally but also as a function of customer satisfaction'.

In the context of this book, the Competing Values Framework (CVF) Model is of the most interest, as it directly contributed to some of the key dimensions of the Denison Organizational Culture Model. This is hardly surprising as Denison and Quinn (two of the originators of the CVF Model) collaborated on research into the complexity of organisational leadership behaviour (Dension *et al.*, 1995), which was extensively based on the use of a CVF concept of measurement, and indeed Denison also refers to the similarity between his own model and that of Quinn and Rohrbaugh (1983) in his book *Corporate Culture and Organizational Effectiveness* (1990: 16). The section that follows for this reason investigates the development of the CVF and underpins its relevance for use in the research described in this book.

Table 4.1 Models to measure organisational effectiveness

Model type	Organisational concept	Organisational focus	Supporting authors
Goal	Rational set of arrangements oriented towards achieving goals	Achievement of outcomes (ends)	Etzioni (1960)
System	Open system (input, transformation, output)	Inputs, resources, processes (means)	Yuchtman and Seashore (1967)
Strategic constituencies	Internal and external constituencies negotiating complex sets of constraints, goals and referents	Responding to expectations of high-powered interest groups that surround the organisation	Connolly *et al.* (1980)
Competing values	Organisation possesses competing values which create many conflicting desired outcomes (goals)	Competing values: 1 Internal vs. external focus 2 Control vs. flexibility 3 Ends vs. means	Quinn and Rohrbaugh (1983)
Ineffectiveness	Organisation viewed as a set of problems/faults	Factors preventing successful performance	Cameron (1984)

Source: Goodman and Pennings (1977), Cameron (1984).

Organisational effectiveness models

The Competing Values Framework (CVF)

One of the better known and possibly the most widely used multi-dimensional models over the past 20 years or so is the Competing Values Framework (CVF), originally developed by Quinn and Rohrbaugh in 1983. Handa and Adas (1996: 343) in their study developing an effectiveness prediction model for construction firms state that the competing values approach 'has also been discussed by Lewin and Minton (1986), Cameron (1986), Quinn and Cameron (1988), Robbins (1990) and Maloney and Federle (1991) and is based on the premise that there is no one criterion for evaluating effectiveness'. Quinn and Rohrbaugh (1983) began by attempting to establish those characteristics that distinguished effective organisations from other, more average companies. They discovered that much of the previous research into this area had established 'fixed' conclusions as to what the main effectiveness criteria were and these were largely incompatible with one another. Highlighting such concerns, Quinn himself (1991: 2) observes, 'Many studies were done in which people set out to measure the characteristics of organisations. These measures were then submitted to a technique called factor analysis.' Factor analysis enabled the production of lists of variables that apparently characterised effective organisations. The difficulty was that in every study carried out, these variables and factors appeared to differ significantly. Handa and Adas (1996: 351) in their very relevant study within the construction industry note, 'Fourteen organisational variables were identified to predict the effectiveness of the construction firm. Analysis of the data from 76 construction firms indicated that only five of the original 14 variables are significant in predicting their level of effectiveness' and these were:

- organisational attitude towards change;
- multiple project handling capability;
- level of planning by management;
- strength of organisational culture;
- level of workers' participation in decision-making.

These conflicting outcomes, together with the inflexible nature of the findings of such studies prevalent at that time, appeared to Quinn and other researchers to present major disadvantages when studying organisations. As a result, Quinn and Rohrbaugh (1983) attempted to classify how various management experts and other researchers actually described organisations based on statistical factor analysis of a comprehensive list of effectiveness indicators. Their results showed that organisational theorists tended to share a cognitive map or an implicit theoretical framework, and from this finding they evolved the Competing Values Framework (CVF) model, adapted by Coffey (2003) and shown in Figure 4.1.

The CVF was evolved after organising, consolidating and integrating multiple effectiveness criteria into three sets of incompatible dimensions. The researchers discovered that there appeared to be two major dimensions underlying the various concepts of effectiveness. These are:

1   Organisational focus (internal emphasis on the well-being and development of people in the organisation to external emphasis on the development of the organisation itself).
2   Different preferences of organisational structure (representing the contrast between stability and control and between flexibility and change).

Use of a multidimensional approach for assessing organisational effectiveness has been discussed and supported by many authors (Keeley, 1984; Cameron, 1986; Ostroff and Schmidt, 1993; Annold *et al.*, 1995) and the conclusion from many of their studies is that due to the variety of different organisational situations and requirements, multidimensional models are the most appropriate for measuring effectiveness. Cameron (1980, 1986) connects these together, as is shown in the Venn diagram in Figure 4.2.

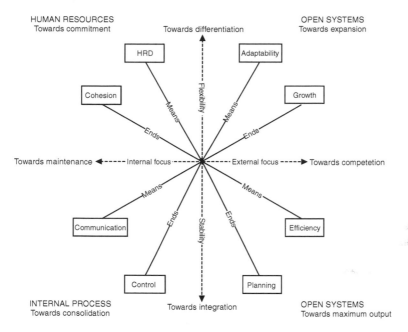

Figure 4.1 The Competing Values Framework model (CVF).
Source: Quinn and Rohrbaugh (1983) adapted by Coffey.

Figure 4.2 Typical differing organisational requirements.
Source: Cameron (1986) adapted by Coffey.

Cameron (1980: 66–80, 1986: 539–553) stated that the guidelines (approaches) for choosing the appropriate criteria are:

- *goal accomplishment* – appropriate when goals are clear and measurable;
- *resource acquisition* – when inputs have a traceable impact on results;
- *internal processes* – appropriate when organisational performance is strongly influenced by specific processes;
- *strategic constituencies* – when powerful stakeholders may significantly benefit or harm the organisation.

Robbins (1990) developed a multidimensional construct model which seemed to be accepted by organisational theorists and several variants, or newer models, were developed during the 1990s that have been used by researchers to examine effectiveness (Bhargava and Sinha, 1992; Ridley and Mendoza, 1993; Jackson, 1999), yet there is still little consensus as to what represents the most valid set of criteria for measuring effectiveness. Goal-based measures of organisational effectiveness had become largely defunct by the end of the 1980s, but they continued to drive the measurement and assessment of corporate performance well into the mid-1990s. However, a growing body of opinion had by the early 1990s, begun to oppose these kinds of measures (Kaplan and Norton, 1992; Brown and Laverick, 1994; Baltes and Parker, 2000). Supplementing this view, it appears to some other authors (McKiernan and Morris, 1994; Hoy and Miskel, 1996) that any valid model of organisational performance needs to be multidimensional in nature in order to adequately measure such aspects of an organisation's performance as:

- system capability;
- system characteristics;
- degree of planning;
- financial performance;
- goal attainment.

In the construction industry, several models developed for measuring performance and effectiveness, or improving quality outcomes, efficiency and business excellence, became popular during the past decade and some of these are reviewed in the following sections.

Total Quality Management

Total Quality Management (TQM) is a holistic and systematic approach to management, based on a set of generic principles that evolved from William Shewhart's pre-Second World War Statistical Process Control. The philosophy of TQM has usually been ascribed to Deming (1986), with a subsequent evolution that encompassed the postwar renaissance of Japanese industry and the widespread quality circles and quality assurance movements in the USA and Europe of the 1960s and 1970s. TQM became the high-fashion management fad of the *post-capitalist* 1980s and 1990s (Drucker, 1994), capturing the imagination of managers both in private firms and the public service (Eicher, 1992). TQM ensures that organisations '*do the right thing*', while the ISO 9000 quality management series is about '*doing things* right', although the two are often confused (Martínez-Lorente *et al.*, 1998). While it is acknowledged that TQM can substantially improve company effectiveness and business excellence (Cassidy, 1996), most organisations find it hard to implement TQM for a variety of reasons (Maital, 1984; Salazar, 1989). According to Tan (1997), among these are:

- large size, diversity and locations of organisations;
- resistance to changes in behaviour, habits and relationships between leaders and employees;
- lack of conviction that TQM works;
- weak organisational performance ethics and challenges;
- reward of individuals rather than team by organisation and intrinsic preference for individual over group accountability;
- most organisations do not understand what quality means and how to measure it.

In the UK construction industry arena, recognising the absence of comprehensive sets of guidelines for implementing TQM, the Construction Industry Institute (CII) commissioned a task force to be responsible for researching how best to implement TQM in the engineering and construction industries, and in 2004 they produced a 'TQM Implementation Roadmap' for use by building and engineering companies. More recent research (Motwani, 2001a, 2001b; Chin and Pun, 2002) has focused on examining the measurement aspects applicable to the operation of existing TQM models, the latter specifically with reference to establishing a framework of critical factors applicable to implementing TQM in Chinese companies and the former to developing a universal model for measuring TQM performance in any sort of business enterprise. Based on this latest research, it appears that the 'measurement' aspect of TQM models is found in the checking of compliance and performance against espoused goals of the model and in the review of achievements against targets by management. While TQM has sometimes previously been mistakenly viewed as a goal-based model (probably due to being seen as a logical evolution from ISO-based quality management systems which are goal-based), modern TQM measurement models such as the Baldridge Award and EFQM have firmly established TQM as a bonafide multidimensional tool. Through the work of researchers such as Motwani (2001b), TQM has made great leaps forward in developing a more comprehensive set of measurement dimensions for use in evaluating organisational effectiveness.

Business excellence models and awards

Evolving out of the interest in ISO quality management systems and the TQM philosophy that emerged in the 1980s, various business excellence models and performance measurement systems have developed. Such models and systems are another way by which organisations 'measure' their effectiveness and some of the major examples of these are described below. According to The UK Department of Trade and Industry's web-based publication *From Quality to Excellence* (2004),

> Traditionally, organisations have always measured performance in some way through financial performance. ... In a successful total quality organisation; performance will be measured by the improvements seen by the customer as well as by the results delivered to other stakeholders, such as the shareholders.
>
> (DTI 'Performance', Chapman, 2009: 1)

The early forerunners to contemporary excellence awards included the Deming Prize, which originated in Japan in 1951, and the Malcolm Baldrige National Quality Award (MBNQA), which was started in 1987 in the USA. Commenting specifically on 'how' these models assist in measuring effectiveness and performance, the DTI publication (DTI 'Excellence', 2004: 1) continues, 'organisations everywhere, of all types and sizes, are under constant pressure to improve their business performance, measure themselves against world class standards and focus their efforts on the customer'.

The EFQM model

To help in this process, many are turning to new total quality models, such as the European Foundation for Quality Management's (EFQM) 'Excellence Model', promoted in the UK as Business Excellence by the British Quality Foundation (BQF). A primary motive for applying for a quality award is not solely to win it, but to compare progress towards self-set excellence targets against other organisations in the field by conducting a self-assessment from which detailed benchmarking feedback is obtained. All of the various awards operating currently are based on the same framework as the European Award. An illustration of the EFQM model adapted from their website in Figure 4.3 shows five of the criteria grouped to form the enablers, covering what the organisation 'does' which cause the other four, grouped into' results', i.e. what an organisation achieves (from the perspective of its major stakeholders).

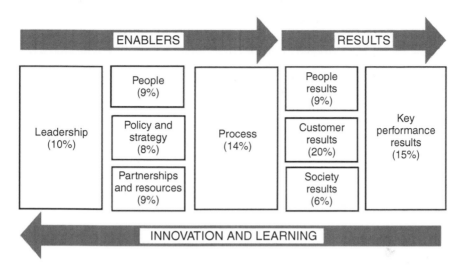

Figure 4.3 The EFQM Excellence model.
Source: EFQM (2009) adapted by Coffey.

It is these 'results' that are of most interest to the concept of organisational effectiveness, as it is these hard metrics which may be used to inform an organisation where it 'is', where it needs 'to be' and what the 'gap' is to be bridged. From Quality to Excellence (DTI 'Excellence', Chapman, 2009: 1) concludes that 'There is significant evidence, including research into organisations that have won national and international excellence awards, of the benefits that can be obtained from following a philosophy of excellence in business'.

However, not all of the feedback on using such models either as a measurement tool or as a change motivator completely extols their benefits to an organisation. The following descriptions from the 'Vanguard Education' website (Seddon, 1998) evoke two fairly common criticisms of business excellence models. The first is related to their inability to deliver the expected or soughtafter improvements to the 'business' of an organisation: 'Many early users of the Model are disenchanted, they haven't seen a rise in quality, excellence or the bottom line' (Seddon, 1998). The other major criticism of such models is their inappropriateness of using scores alone to improve performance or give an accurate analysis of what a company must do to improve. This is how a director of an engineering organisation spoke of her experience with implementing the Baldridge Excellence Model (BEM) self-assessment:

> My advice to anyone using the BEM is start with processes ... we didn't get the quantum leap we were hoping for because *we went down the*

scoring route [emphasis by author] ... we should have started with a definition of our work processes and focused on ... how we deliver what matters to our customers ... *change should not start with comparison to a model, it should start with a thorough understanding of the 'what and why' of current performance.*

(Seddon, 1998)

The balanced scorecard

Aware of the disadvantages and deficiencies of performance measurement approaches that had been used in the late 1970s and early 1980s and following on from the research of Johnson and Kaplan (1987), Keegan *et al.* (1989) challenged [in their view] the continued use of such inappropriate measures in an article entitled 'Are your performance measures obsolete?' Kaplan and Norton (1992) further developed Keegan's views and produced their own 'balanced scorecard' (BSC), an example of which is shown in Figure 4.4.

The BSC approach established exactly what companies need to measure, other than just financial success factors. Businesses worldwide accept

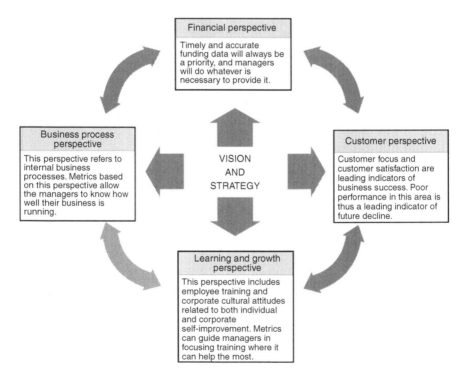

Figure 4.4 The Balanced Score Card model.
Source: Kalpan and Norton (1992) adapted by Coffey.

the BSC as a virtually self-contained management system and also as a powerful performance measurement system that provides feedback on internal business processes, as well as external results. These real-time data enable organisations to continuously improve strategic performance and business outcomes. Kaplan and Norton (1996: 7) describe the innovations of the balanced scorecard system as follows:

> The balanced scorecard retains traditional financial measures. But financial measures tell the story of past events, an adequate story for industrial age companies for which investments in long-term capabilities and customer relationships were not critical for success. These financial measures are inadequate, however, for guiding and evaluating the journey that information age companies must make to create future value through investment in customers, suppliers, employees, processes, technology, and innovation.

Arthur M. Schneiderman, the former vice-president of Quality and Productivity Improvement at Analog Devices, Inc., has written two contrasting view articles on the balanced scorecard approach. One of these articles (Schneiderman, 2009a) sets out the historical success of applying the balanced scorecard at Analog Devices between 1986 and 1991, entitled 'The first balanced scorecard'. In another article entitled 'Why balanced scorecards fail!', Schneiderman (1999) offers the reader six reasons most often responsible (in his personal view) for the failure of the balanced scorecard approach, namely:

1 independent (i.e. non-financial) variables are incorrectly defined as primary drivers of future stakeholder satisfaction;
2 poorly defined metrics;
3 improvement goals are not based on stakeholder requirements, fundamental process limits and improvement process capabilities, rather they are negotiated;
4 high-level goals are not sufficiently broken down and deployed to the sub-process level where actual improvement activities reside;
5 improvement systems used are generally not state-of-the-art;
6 there is no quantitative linkage between non-financial and expected financial results.

Although the use of the BSC has become widespread in modern business organisations, there are many critics of the model who have identified the following shortcomings:

• It lacks a 'competitiveness' dimension (Neely *et al.*, 1995).
• It does not adequately measure the human resources perspective/ employee satisfaction/supplier performance/product or service quality/

environmental community perspective (Maisel, 1992; Brown, 1996; Ewing and Lundahl, 1996; Lingle and Schieman, 1996).

- It does not reflect the different dimensions of performance most relevant to success such as internal and external customer perspectives or the generic strategic objectives of cost, quality and time (Kennerley and Neely, 2002).
- It lacks a sound foundation and needs to ensure that the basic structural requisites include processes to guarantee that the right things go on the scorecard, with properly defined metrics and rational, time-based goals (Schneiderman, 2009b).

In two more recent criticisms, which have particular significance in the construction industry context, Bourne (2001: 48) reports:

> while some studies show a link between the use of the scorecard and an improvement in share price, many businesses still feel underwhelmed. The main criticisms of the scorecard concentrate on its absence of a people perspective, that it doesn't take into account the fact that businesses are subject to regulation and the lack of any environmental or community issues in its makeup.

Brignall (2003: 1) in an unpublished paper, alluding to these last two factors, states that 'however, neither new financial metrics nor the BSC itself cater for the needs of all significant organisational stakeholders. Two notable omissions are the environment and social matters, which are currently enjoying a resurgence of public interest.' Despite such criticisms, the BSC is still a hugely popular performance measurement tool and at least offers the advantage of being adaptable to individual organisational requirements, as Bourne (2001: 50) further remarks: 'it will survive because there are too many advantages to having a multi-dimensional framework for measuring performance.'

The performance prism

One of the most recent models to emerge is the performance prism developed by Andy Neely of Cranfield School of Management, UK and Chris Adams of Anderson Consulting. Neely and Adams (2001: 7) describe the model as follows: 'The Performance Prism has five facets – the top and bottom facets are Stakeholder Satisfaction and Stakeholder Contribution respectively. The three side facets are Strategies, Processes and Capabilities.' This three-dimensional model is gaining some popularity and has been applied to a number of organisations and conditions in order to thoroughly test its applicability in the field. A schematic diagram of the performance prism is shown in Figure 4.5.

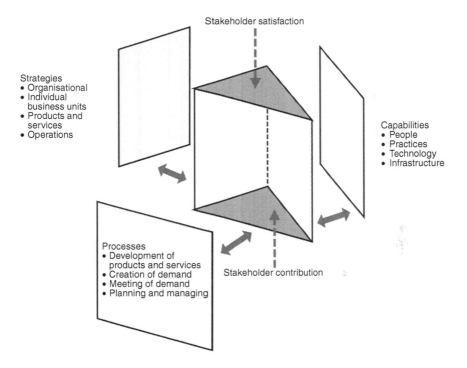

Figure 4.5 The Performance Prism.
Source: Neely and Adams (2001) adapted by Coffey.

Other useful tools briefly described

Management by Objectives

Management by Objectives (MBO) appeared as a 'new' management philosophy during the 1950s and although George Odiorne is usually attributed with having produced the first book dedicated to MBO, the theory of setting up a system of managing businesses by establishing and monitoring organisational members' performance against a set of commonly agreed objectives was first outlined by Peter Drucker in 1954 in his book *The Practice of Management.* According to Kotelnikov (2004), Drucker advocated that managers should avoid 'the activity trap', i.e. of becoming so involved in their day-to-day activities that they forgot their main purpose or objectives. One of the concepts of MBO was that instead of just a few top managers, all managers of a firm should participate in the strategic planning process in order to improve the feasibility of the plan. Another major concept was that subordinates, together with their managers, sat down and agreed on a set of achievable objectives for the subordinate's job. The agreement covered a time frame in which the objectives were to be achieved, a means of measuring how successful they were, whether the subordinate had the sole responsibility for achieving these objectives or whether responsibility was shared with others. The theory was that having agreed on all of the relevant constituent details, the manager could then let the subordinates 'get on with it', only meeting with them from time to time to check on their progress (Odiorne, 1965). Following the publication in 1965 of Ordione's book, *Management by Objectives: A System of Managerial Leadership*, MBO soon became even more popular and was even considered by some as a 'panacea for business management to carry organisations through into the sixties' (Kotelnikov, 2004). The assumption of the MBO system was that organisations would eventually become more effective if members focused on outcomes (achieving results) rather than processes (doing things). MBO principles are:

- cascading of organisational goals and objectives;
- specific objectives for each member;
- participative decision making;
- explicit time period;
- performance evaluation and feedback.

MBO also introduced the 'SMART' method for checking the validity of objectives, which had to be:

- **S**pecific
- **M**easurable
- **A**chievable
- **R**ealistic, and
- **T**ime-related.

Kotelnikov (2004) states:

> unfortunately Drucker never wrote a how-to-do-it book. Consequently, those who have been interested in the idea have seen what they wanted to see and used whatever parts seemed useful to them. This has led to a rather checkered history of acceptance and utilization of MBO; it has been praised, criticized, offered as the magic solution to management problems, or spurned as a waste of time.
>
> (Kotelnikov, 2004)

Tarrant (1976: 82) notes that Drucker began to downplay the significance of this organisational management method, stating, 'It's just another tool. It is not the great cure for management inefficiency ... Management by Objectives works if you know the objectives, 90 per cent of the time you don't.' The focus of MBO is now changed; from one based on an individual manager to one on the operations of the total organisation. It has come

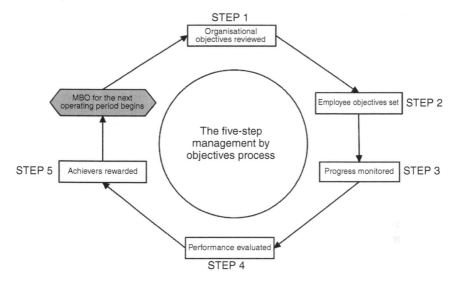

Figure 4.6 The Management By Objectives (MBO) model.
Source: Drucker (1954) adapted by Kotelnokov (2004) and Coffey.

to be used to improve the major steps in the process of carrying out the work, and effectiveness in this system is therefore measured against a set of mutually agreed objectives. Figure 4.6 shows an adapted five-step version of Drucker's MBO process model.

Six Sigma

Linderman *et al.* (2003: 193) define the concepts and principles underlying Six Sigma as follows:

> [It] is an organized and systematic method for strategic process improvement and new product and service development that relies on statistical methods and the scientific method to make dramatic reductions in customer defined defect rates.

The goal of Six Sigma is 'bottom-line' financial improvement and, once implemented, the following benefits have been recorded as resulting:

- productivity increases;
- cycle time reduction;
- higher throughput;
- reduced defects;
- high levels of outgoing quality;
- standardised improvement methodology across the organisation;

- a set of techniques and tools to simplify improvement efforts;
- greater customer satisfaction;
- dramatic improvement in the 'bottom line'.

Statistically the Greek letter 'σ' (sigma) is used to denote the standard deviation of a set of data. Reporting sigma, together with the mean value, gives an indication of how all the data points vary from the mean. This is crucial, as reporting the mean value alone can often be misleading. Fellows and Liu (2003) have demonstrated this by describing an analogy of someone having two feet in a hot oven and the head in a bucket of ice, but on *average* the person was doing 'OK'. Under the Six Sigma definition, 3.4 defects per million or 99.99966 per cent represents good quality (although zero defects is the goal) and the philosophy is that an organisation operating its product delivery and services to this measure will drive a firm towards achieving higher levels of customer satisfaction and reducing operational costs. The 'Six Sigma Partnering LLC' (2004) website states that,

> Six Sigma is a robust continuous improvement strategy and process that includes cultural methodologies such as Total Quality Management (TQM), process control strategies such as Statistical Process Control (SPC) and other important statistical tools. It uses a structured systems approach to problem solving and strongly links initial improvement goal targets to bottom-line results.

The framework used to achieve Six Sigma goals is known as DMAIC (Define, Measure, Analyse, Improve and Control). Figure 4.7 demonstrates how DMAIC sits around the Deming PDCA (Plan, Do, Check, Act) model, confirming its close association with TQM philosophy.

In its formative years, DMAIC was practised and perfected on performance improvement initiatives directed at existing processes that resulted in manufacturing defects. Today, the methodology is used for many business processes that fail to meet customer requirements. According to Harry and Schroeder (2000), the DMAIC approach involves:

1. Defining and understanding the problem being addressed by identifying the critical customer requirements and key factors affecting the process output.
2. Measuring relevant data to the problem primarily through Six Sigma metrics.
3. Analysing, using statistical quality control tools, the production or business process associated with the problem to identify the root causes.
4. Improving the process using alternatives derived in the analysis phase.
5. Controlling and monitoring the process using statistical process control to sustain the gains and improvements.

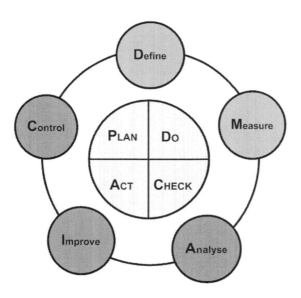

Figure 4.7 The combined DMAIC and PDCA model.
Source: FYCC (2008) adapted by Coffey.

Breyfogle *et al.* (2001) quote Tom Pyzdek, the Six Sigma 'guru', as saying:

> Six Sigma is such a drastic extension of the old idea of statistical quality control as to be an entirely different subject ... in short, [it] is ... an entirely new way to manage an organisation ... Six Sigma is not primarily a technical program; it's a management program.
>
> (Breyfogle *et al.*, 2001: 15)

However, despite its popularity in process-driven and production organisations, many authors have criticised the relevance of Six Sigma since, unlike TQM, it places pre-eminence on quality process control at the expense of all other business aspects (Harry and Schroeder, 2000; Pande *et al.*, 2000; Breyfogle *et al.*, 2001).

Benchmarking

This was a concept developed in the USA in the 1970s; however, the underlying theory has been in existence for considerably longer. Emanating originally from the studies on the scientific methods of work organisation performed by Frederick Taylor in the late nineteenth century, benchmarking was originally developed by companies operating in an industrial environment. As defined by the US Department of Energy (DOE)'s Benchmarking Clearinghouse, 'benchmarking is the process of identifying, understanding, and adapting outstanding practices and processes from organisations anywhere in the world to help your organisation improve its performance' (USDOE, 2004). The Scottish Centre for Facilities Management (SCFM) has a slightly different definition: 'benchmarking is a performance measurement tool used in conjunction with improvement initiatives to measure the operating performance of companies on a comparative basis and identifies "best practice"' (SCFM, 2004). In recent years, organisations such as government agencies, hospitals and universities have also discovered the value of benchmarking and are applying it to improve their processes and systems. Benchmarking is a tool for improving performance by learning from best practices and understanding the processes by which they are achieved. The successful application of benchmarking involves four basic steps:

1   First, understand in detail your own processes.
2   Next, analyse the processes of others.
3   Then, compare your own performance with that of others analysed.
4   Finally, implement the steps necessary to close the performance gap.

According to a recent report by the University of Technology, Sydney (2002: 1): 'Benchmarking is best undertaken for processes that are well understood, have the capacity for improvement, and will potentially benefit from external study and innovation.'

Other commercially developed effectiveness measures

While this review of measures of effectiveness has largely concentrated on instruments used by researchers in an academic context, it should be borne in mind that many very good commercially produced surveys or inventory tools (some based on models originally developed to support academic organisational research) were used to measure and evaluate performance, and Green and Henderson (2000) reviewed nine such effectiveness surveys (see Table 4.2).

Some of the instruments listed were developed and used by researchers to investigate concepts such as team working, team problem-solving and conflict management (Cooke and Lafferty, 1986; Green, 1992; Hallam and Campbell, 1997; Anderson and West, 1998) organisational satisfaction and leadership competency (McCarthy *et al.*, 1997), organisational culture and effectiveness (Cooke and Lafferty, 1986; Cooke and Rousseau, 1988; Anderson and West, 1998), employee satisfaction and attitudes (Folkman, 1998), and nearly all of them are widely used by management consultants to diagnose organisational weaknesses and drive organisational change. In relation to the subject matter of this chapter (i.e. the link between organisational effectiveness and organisational culture), with the exception of the two Human Synergistics International surveys (OCI and OEI), none of

Table 4.2 Nine ways to evaluate team-based effectiveness

Name of instrument	Theory
Campbell-Hallam Team Development Survey (TDS) NCS Pearson. USA	Designed to identify team strengths and weaknesses and stimulate discussion about critical team issues
Group Styles Inventory (GSI) Human Synergistics, USA	Designed to be administered after a team tackles and attempts to solve a particular problem. It measures the patterns of behaviour that emerge when people work together toward a solution
Organisational Analysis Survey (OAS) Novations Group, Inc., USA	Measures employee opinions and attitudes which allow management to see how employees feel and compare it to how management thinks employees should feel
Employee Opinion Survey (EOS) QPT Associates, USA	Designed to elicit opinions employees have about their workplace
Winning Manufacturing Diagnostic (WMD) Thompkins Association, Inc., USA	May be used to evaluate or determine the continuous improvement priorities of a manufacturing firm
Organisational Cultural Inventory (OCI) Human Synergistics, USA	To assess thinking and behavioral styles. The scores identify the type of corporate culture
Organisational Effectiveness Inventory (OEI) Human Synergistics, USA	Seeks to identify structural features, processes, practices and outcomes that affect service, quality and productivity. Some outcomes that are assessed include service quality, team work and coordination
Team Climate Inventory (TCI) Hanover House, UK	Focuses on four characteristics of group innovation and effectiveness (vision, participative safety, task orientation and support for innovation)
Quality Performance Teamwork Inventory (GOTEAMS) QPT Associates, USA	Designed to measure and assess each feature and stage within the 'GOTEAMS framework' which includes seven features identified as having a bearing on implementing successful team-based organisations

Source: Green and Henderson (2000), adapted by Coffey (2003).

these instruments was particularly appropriate to measure the concept of effectiveness, as many of them are either 360° feedback-type surveys designed to provide employee views on management and leadership styles, or they are team/group measures of strengths/weaknesses or cohesion. The most well-validated instruments from the list above are the OCI and OEI, which are described by the 'authors' as measuring organisational culture and effectiveness respectively. Closer scrutiny reveals that OCI is designed to profile cultures into patterns or styles of behavioural norms and expectations based on asking questions which make respondents 'identify the extent to which a variety of behaviours are encouraged and required of people within the organisation' (HSI, 2004). The OEI (which is designed to be used in conjunction with the OCI) focuses more on identifying the causal factors that influence culture and its outcomes, and thus help management to strategically intervene to improve organisational processes and performance. The real value of these instruments is to assist individual companies seeking to establish deficiencies in their corporate cultures and thus develop change programmes to improve their bottom-line effectiveness. While the OCI is undoubtedly a very useful tool for establishing the behaviours within corporations that may contribute towards successful business outcomes, in does not appear to actually 'measure' the strength of organisational culture. It does measure 'styles' or modes of thinking, behaving and interacting in an organisation with a view to diagnosing behavioural characteristics that may provide a basis for change programmes. Combined with the OEI, it produces an integrated measure of behavioural norms and effective behaviour and as such is difficult to use to correlate against external performance measures. The OCI does not, according to the seminal literature on the instrument previously mentioned, appear to have been put to the same validity and reliability testing as the Denison model, although it may well have been used on a similar number of recipient organisations as the DOCS. The traits of this model appeared to be less relevant to a technological organisation such as a construction company than the DOCS. The measurement dimensions of the OCI focus on psychological or philosophical aspects of an organisation as viewed by internal participants, whereas DOCS focuses on measuring the management practices of organisations that specifically impact on culture, thus being able to 'snapshot' an organisation's progress towards becoming a high-performance cultural entity, when compared to a central database of thousands of assessments using the same DOCS instrument. For these reasons, in the construction-related research study described later in this book, it was decided not to use the OCI to test the organisational culture and performance link.

In recent years, multidimensional models have been widely advocated in the literature most closely related to performance assessed on a perceptual basis (Preston and Spienza, 1990; Doyle, 1994; Donaldson and Preston, 1995), and some researchers see the next logical development (i.e. a multi-stakeholder assessment approach) as being the next most favourable construct (Tsui, 1981, 1990; Wilderom *et al.*, 2000). As the research focus in this book examines the links between organisational culture and performance and espouses to use measures of effectiveness that are designed to be superior to the financial measures used previously, particularly where a group of different organisations are being assessed, the appropriate instrument used to assess effectiveness is required to allow comparisons between effective and non-effective organisations in sufficient depth to identify and measure those detailed aspects of performance which contribute to the overall success of the organisation. This is particularly important when such measures are being investigated in conjunction with other measures of organisational strength and weakness such as culture. Based on the various discussions covered in this chapter, the Denison model (which specifically measures organisation cultural strength) was used in conjunction with an independent and separate external measure of effectiveness (i.e. the Performance Assessment Scoring System (PASS)) to operationalise the concepts of organisational culture and organisational effectiveness in the research described in Chapters 6, 7 and 8 of this book.

Chapter summary

This chapter has examined the various theories, models and measures of business performance and organisational effectiveness, and has attempted to draw some comparisons between them. The chapter ended by linking the concepts encompassed in one measure of organisational culture and the effectiveness and critical success measures used in one specialised sector of the Hong Kong construction industry so that the reader can begin to understand their connection described in later chapters of the book.

5   Organisational culture and effectiveness

Investigating the link between them

Chapter introduction

This chapter focuses on the development of research into the link between organisational culture and effectiveness, and examines the different periods in the history of this thematic development. The chapter looks in depth at the Denison Organisational Culture Survey which was chosen to support the research study among construction companies in Hong Kong, the methodology and results of which are described later in the book. The chapter ends with a view of the relatively few studies into culture so far carried out specifically in the construction industry.

Early studies: the 'Budding Stage' (1920s to 1970s)

In her review of the literature on studies attempting to link organisational culture and effectiveness, Liu (2003: 1) states that 'The purpose of studying organisational culture is many fold; but the principle purpose is to find out how it effects the organisational performance with a view to improving the performance effectiveness.' Kotter and Heskett (1992: 8–9) state that 'Cultures can have powerful consequences, especially when they are strong ... these cultures can exert a powerful effect on individuals and on performance, especially in a competitive environment', and according to Wilderom *et al.* (2000: 193), 'Research on the link between organisational culture and performance/effectiveness has increased substantially during the last decade'. As the main research question examined in this book revolves around this relationship between the culture and performance of businesses (specifically Hong Kong construction businesses) the following section reviews the literature documenting other research in this area.

According to various authors (Vesson, 1993; Wilderom *et al.*, 2000; Liu, 2003), Organisational Culture–Organisational Effectiveness (OC–OE) linkage studies have evolved through three distinct historical stages, which are:

1   Budding Stage
2   Promulgation Stage
3   Testing Stage.

These stages will now be examined in detail and their contribution to the overall extant literature on the OC–OE linkage discussed. The earliest research studies (i.e. the Budding Stage) examining the possibility that some form of linkage between the 'socialisation' which occurred in organisations and the way in which organisations ultimately performed can initially be traced back to the 'Hawthorne Studies' conducted between 1924 and 1933 at the Western Electric Hawthorne Works in Chicago (Wilderom *et al.*, 2000: 194) by Professor Elton Mayo. In the second phase of the studies, Mayo and his team examined the relationship between workplace environment and productivity, and their results led to the formation of a fundamental concept that the workplace is essentially a social environment, and, depending on the extent of socialisation that occurs and the level of team building which has been achieved, productivity is positively affected. It was from such roots that the 'Human Relations School' of management thinking evolved, which held sacred, among other values, that organisations should seek to improve their internal and external human relations not just because of the effect that such improvement could have on productivity but with a long-term view to producing a balanced attitude towards also satisfying the personal and social needs of workers (Roethlisberger and Dickson, 1939). Even more significant were the findings from the studies carried out by the Tavistock Institute of Human Relations at the Glacier Metal Works, starting in 1948. The original experiment was established to investigate the many and varied problems in industrial organisations, which had not been satisfactorily answered up until that time. One of the most baffling questions was related to the syndrome of demotivation and alienation towards employers that workers exhibited after they had been in the job, sometimes for only a short time. This dissatisfaction often culminated in workers organising themselves into groups in order to pressurise employers to pay them higher wages. The Glacier project tested the 'theory of equitable payment' developed by Elliot Jaques, a well-known psychologist in the 1940s. Jaques (1951) was mainly concerned with the social interactions among employees and the subsequent regularity and stability in work productivity and performance that resulted from such socialisation. While the Glacier model did appear to show that the concept of a clearly understood wage differential acted as a motivator for better performance, Jaques' findings did not predominantly find any congruency between organisational culture, organisational structure and organisational environment. On the contrary, what his research did appear to reveal was that if culture was not aligned with an organisation's structure or social (i.e. work) environment, a barrier to effective productivity (i.e. 'performance') could result (Wilderom *et al.*, 2000). Studies which followed that of Jaques during the decade between 1960 and 1970 that focused on organisational culture did not specifically address the link with effective performance or business, although Pfiffner and Sherwood (1960: 250) note that as every organisation evolves, 'certain patterns of conducts and beliefs ... become the value system of the organisational members, and it is within this context that all members are expected to operate'. Pfiffner and Sherwood proposed (but did not actually investigate) that there could be a link between the effectiveness and culture of an organisation (Wilderom *et al.*, 2000). Silverzweig and Allen (1976) carried out one of the earliest studies on normative systems in organisations (i.e. a set of expected behaviours or '*norms*' that are generally supported by people in the organisation) which produced results suggesting that cultural interventions were significantly related to an organisation's performance. Their normative system model, shown in simplified format in Figure 5.1, is based on four interlocking circles, each representing a separate phase in the change management process.

Based on their study of eight companies, Silverzweig and Allen (1976) discovered that in six out of eight cases, organisations directly altering their cultures following periods of sustained business losses were able to substantially improve (and maintain) their performance over long periods

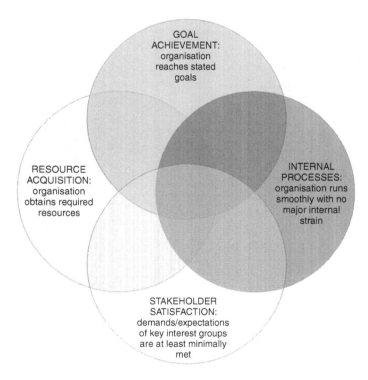

Figure 5.1 The Normative Systems model.
Source: Silverzweig and Allen (1976) adapted by Coffey (2003).

of time as a result of such culture change. It should be noted however that this particular study did not end up proving a direct relationship between culture and performance (as the performance measures used differed depending on the organisation being studied and were thus not comparable). Wilderom *et al.* by way of criticism (2000: 195) state that 'Their study had more value for research on the effects of cultural interventions than for exploration of the relationship between culture and performance'. These authors do conclude however that such results may well have provided impetus for others to study the C–P linkage in future.

More recent studies: the 'Promulgation Stage'(1980s)

By the end of the 1960s the interest of researchers into the possible catalysts for business performance enhancement in the United States had moved away from attempting to prove the efficacy of the C–P linkage and by the early 1970s attention had focused excitedly on widespread investigations into the reasons behind the worldwide success of Japanese manufacturing organisations. This movement was led by Ouchi, who stated that the business success of Japanese organisations was largely due to their more humanistic concern for employees and support for consensus-based decision-making. This concept of performance being linked to commitment and workers and employers sharing the same unified vision is a pivotal theme found in Ouchi's definitive research collaborations in the late 1970s (Ouchi and Jaeger, 1978; Ouchi and Johnson, 1978). This work by Ouchi and his colleagues therefore provides the first 'direct' linkage between the strength of organisational cultures and differing degrees of effective performance. In 1981, Ouchi released *Theory Z: How American Business can Meet the Japanese Challenge*, which further developed his earlier collaborations. Martin (2002), using these three works, compares the dimensions of three organisational types to show how American organisations may have evolved as a result of the effects of such dimensions on their organisational development. Ouchi's views were supported by Pascal and Athos (1981) who, in the same year that *Theory Z* was published, released their own book *The Art of Japanese Management*, which extended and popularised this area of investigation. In their book, the authors described a study of 30 Japanese companies, focusing in particular on 'Matsushita' as an example and contrasting their management to that of several US companies. These authors concluded that management and decision-making in Western businesses was essentially top-down and hierarchical, being largely modelled on military command concepts, with levels of authority, unity of command, line and staff functions, and so on. Their resultant analysis of American and Japanese business management styles postulates that Japanese companies were altogether superior to American firms in terms of skills, staff-quality, shared values and management style (i.e. the 'soft Ss' (style, skills and staff)) along with the others. They stress the importance of organisational values, particularly finding a balance between the scientific operation of production and the management art of dealing with the people involved, and it is this link that in their view is responsible for creating an excellent corporation. Although Pascale and Athos (1981) first introduced the concept of 'Seven Ss', the actual model was born at a series of discussions between themselves and the McKinsey consultants, Peters and Waterman, in 1978; the model developed a framework by which to measure a company's 'excellence'. Peters and Waterman went onto publish the 'Seven-S-Model' (better known as 'McKinsey 7-S'), first as an article (1980) and later in their book *In Search of Excellence* (1982). The 'Seven-S-Model is depicted in Figure 5.2.

The authors concluded, based on an in-depth study of 21 out of 43 'excellent' firms drawn from an original population of 62, that there is a strong link between the culture of a company and business performance. One of the main conclusions of their work was that if companies replaced outmoded prescriptive scientific rationalised management styles with a more flexible, humanised and adaptive model, superior performance would be facilitated. This view was to set the mode for subsequent studies, as interest was now firmly rooted in the benefits of driving organisational change for the good by modifying the existing company cultures. The following quote from the authors captures this hypothesis:

> Without exception, the dominance and coherence of culture proved to be an essential quality of the excellent companies. However, the stronger

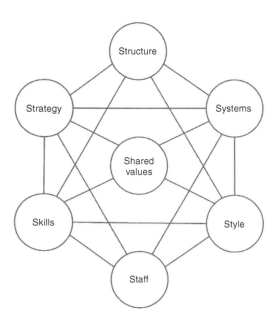

Figure 5.2 The Seven-S-model.
Source: Peters and Waterman (1982) adapted by Coffey.

the culture and the more it was directed toward the market place, the less need there was for policy manuals, organisational charts, or detailed procedures and rules.

(Peters and Waterman, 1982: 79)

According to Carroll (1983), two further facts emanating from that study need to be noted:

- That a large number of the firms identified as 'excellent' in the book subsequently turned out to actually prove themselves to be disappointing performers; some even went out of business.
- Some of the poorer performers also appeared to have 'strong' cultures, but unlike Japanese manufacturers, their cultures were generally dysfunctional, being focused on 'numbers' and 'internal politics' rather than on 'product quality', 'employees who make and sell the product' and externally on 'customer satisfaction'.

Also in 1982, a piece of work arrived on the scene which was to unlock an entirely new direction for business research and the interest focus of managers worldwide, namely *Corporate Cultures* by Deal and Kennedy (1982). This study hardened the focus of the previous 'softer' popular studies in that it argued that companies evolve and develop distinctly different types of cultures and these have a measurable and direct impact on both organisational strategy and performance. They further postulate that a positive 'nurturing' element directed towards corporate human resources within such cultures is absolutely critical to the company's ultimate success in their market sector. In setting down the 'importance of culture', the authors have this to say:

> Companies that have cultivated their individual identities by shaping values, making heroes, spelling out rites and rituals, and acknowledging the cultural network have an edge. Those corporations have values and beliefs to pass along – not just products. They have stories to tell – not just profits to make. They have heroes whom managers and workers can emulate – not just faceless bureaucrats. In short, they are human institutions that provide practical meaning for people both on and off the job. We think that people are a company's greatest resource, and the way to manage them is not directly by computer reports, but by the subtle cues of culture. A strong culture is a powerful lever for guiding behaviour; it helps employees do their jobs a little better, especially in two ways:
>
> - a strong culture is a system of informal rules that spells out how people are to behave most of the time; (and),

- • a strong culture enables people to feel better about what they do, so they are more likely to work harder.

(Deal and Kennedy, 1982: 15)

The 'Testing Stage' (1980s to the present)

Following on from Deal and Kennedy's work came several research publications which criticised, or at least challenged, the C–P link, and one of the most scathing was that of Carroll (1983) which set about questioning the methods employed by Peters and Waterman, stating that they had not compared the values of other 'non-excellent' companies to see if they differed from those companies researched for the book. Carroll also noted that several of the 36 companies studied suffered serious financial downturns a few years after the study was completed. This view was further supported by popular press commentaries (*Business Week*, 1984) that were based on the apparent declining financial performance of at least 10 of the 43 'superior' companies described by Peters and Waterman, many (including Amdahl and Data General) of which fell on hard times soon after publication. It appears that Carroll's view may have been somewhat prophetic, as in a more recent article in the online version of *Business Week*, Peters was quoted as saying, 'for what it's worth, okay, I confess: we faked the data' (Byrne, 2001). The article continues, 'For years, many assumed that the authors employed rigorous research and stringent financial screens to identify "excellent" companies.' It now appears that Peters and Waterman 'simply asked their McKinsey colleagues and other "smart people" for the names of companies doing "cool work" '. Reynolds (1986) further challenged the results of Peters and Waterman by establishing little or no difference between responses to an employee culture questionnaire from 'excellent' and less successful companies. This result appeared to be supported by the work of Hitt and Ireland (1987) who studied 185 companies listed in the *Fortune 1000* list (14 of these firms were in Peters and Waterman's original list of superior performers). These authors found that only three of the 'excellent' firms performed better than the average of the general sample and several from the general sample outperformed all excellent firms.

Based on such results, support for the C–P link was by the end of the 1980s under very serious challenge. Two further studies carried out towards the end of the decade, while not entirely rejecting that such linkage existed, served to reinforce this challenge (Saffold, 1988; Siehl and Martin, 1990). Saffold (1988) found that it was not possible to view the relationship in terms of a simple model as based on the knowledge which scholars had developed of culture's role in organisational analysis; a more complex framework needed to be developed by researchers for the future. Saffold's (1988) study contained a comprehensive and critical review of the literature on culture–performance linkage studies, and he highlights the perceived weaknesses in these studies and proposes various areas for development and improvement in future studies. Siehl and Martin (1990) also criticised the approach taken in some of these previous studies with regard to both content and methodology and, while acknowledging the possibility of a link between the two, stated that 'the promise of a link between organisational culture and financial performance is empirically unsubstantiated, perhaps impossible to substantiate'.

Organisational culture: performance link studies (1990 to 2003)

Research studies have over the past decade continued to investigate the linkage between organisational culture and corporate performance and some have established evidence of such a link, thus concluding that it does indeed exist (Denison, 1990; Gordon and DiTomaso, 1992; Kotter and Heskett, 1992; Petty *et al.*, 1995; Wilderom and Van den Berg, 1998). Wilderom *et al.* (2000: 194–209) conclude that 'most of the empirical studies reported in the 1990s had important conceptual and methodological weaknesses' and that 'the research evidence regarding the claimed predictive effect of organisational culture on performance/effectiveness appears to be there, but not very convincingly so'. However, other critical reviews of the methodologies and findings used in such research challenge such conclusions. Lim (1995) reviewed the methodologies and findings of several hitherto notable research studies and found that, at best, the culture–performance link remained unclear. He also criticised technical aspects of the studies, such as the potential bias due to moderator variables, and advocated that such fundamental methodological issues should be more comprehensively dealt with and conceptual applications strengthened in future research. Lim states (1995: 20): 'As it stands, the present examination does not seem to indicate a relationship between culture and the short-term performance of organizations, much less show a causal relationship between culture and performance.' Lim continues in the same paragraph, conceding with regard to the usefulness of studying organisational culture: 'the most important contribution of "culture" towards the understanding of organisations appears to be as a descriptive and explanatory tool, rather than a predictive one.' In a more positive vein, Wilderom *et al.* (2000: 201) state: 'Nevertheless, the great intuitive appeal of the C–P linkage, the preliminary evidence found so far and the many research challenges involved in obtaining the evidence give some reason to still believe in this link.' Kilmann and Saxton (1985) have proposed that a methodology measuring the perceived 'culture gap' between *actual* and *desirable* states of an organisation provides a way forward for research linking C–P. Their idea has been taken up by other researchers (Wilderom and Van den Berg, 1998; O'Reilly *et al.*, 1991), although so far there has been no strong empirical evidence to prove the existence or influence of an organisational culture gap on effectiveness. According to Wilderom *et al.* (2000: 203), 'it forms in our view, a fruitful basis for more refined future C–P research'. Although the research study described later in this book does not actually utilise this methodology to examine the C–P link, it does make some pertinent observations in the final chapter on the culture gap (or misalignment of 'walk and talk') derived from limited use of a qualitative methodology described in Chapter 7, applied to support the quantitative findings of the survey and the subsequent statistical analysis. Wilderom *et al.* (2000: 194–209) provide a thorough review of 10 major empirical C–P link studies in the context of all listed studies. In order to update the current situation reported by these authors, three further recent studies have been added to the overall review of C–P linkage research and although two are still somewhat exploratory, their early results are promising and validate them being launched. First, Fey and Denison's (2000, 2003) study on OC factors contributing to the success of US companies based in Russia concluded that adaptability and involvement were the two key traits. Next, Zhang and Liu's (2006) pilot study on organisational culture types prevalent in Chinese construction firms has yielded some promising early data that suggest 'hierarchy' and 'clan' type cultures are stronger than 'market' and 'adhocracy' types. Finally, Wilderom and van den Berg's (2004) study of Netherlands financial institutions, in which they tested five dimensions of organisational culture, has demonstrated a need for more investigation of other dimensions of culture, and posits that maybe a new comparative model with additional dimensions and measures is required to strengthen the accurate comparison of cultures between organisations. A summary of the review is shown in Table 5.1 which this author has extended to also include the recent studies described above.

By way of a final word on this topic, a recent book on organisational behaviour opines: 'No single approach to evaluation of effectiveness is appropriate in all circumstances or for all organisational types' (Kreitner and Kinicki, 2001: 631–632).

Investigation of the link between organisational culture and organisational effectiveness in the context of this book

The preceding chapters of this book have developed a focus on investigating the link that has been established within the literature between organisational culture and organisational effectiveness. As has been described extensively in the preceding sections of the book, debate still continues among researchers on the best single methodology (or multi-methodology) and the most appropriate instrument(s) with which to operationalise and measure the concept of the culture of an organisation. Scholl (2003: 1) observes that most definitions of culture fall into one of two categories: they are either based on an *outcomes* or *process* perspective. From the previous detailed investigation of the wide range of definitions of 'culture' related specifically to organisations, it is clear that there are both 'outcome-based' and 'process-based' definitions in existence. Deal And Kennedy (1982)

Table 5.1 OC–performance link studies conducted since 1990

Reference	Organisational culture dimensions	Performance measure	Organisations involved	Respondents involved	Evidence for the culture–performance link
Denison (1990)	(a) involvement (b) consistency (c) adaptability (d) mission	Average over 6 years of: (a) return on sales (b) return on investment (c) income/sales ratio (d) income/investment ratio	34 large US firms from 25 different industries	43,747 employees within 6,671 work groups	1 Involvement is positively related to short- and long-term performance. 2 Consistency is positively related to short-term performance, but negatively related to long-term performance.
Rousseau (1990)	(a) team- or satisfaction-oriented norms, (b) security-oriented norms	Amount of money raised by community	32 large units of a US nationwide voluntary service organisation	263 paid staff members	Little emphasis on security-oriented norms is significantly related to high performance.
Calori and Sarnin (1991)	work-related values (12 dimensions) and management practices (17 dimensions)/culture strength	Average over 3 years of: (a) return on investment (b) return on sales (c) growth	5 French firms with a single business, in mature industries pursuing a differentiation strategy	280 managers and employees, excluding front-line workers	1 Many values and their corresponding management practices were related to company growth. 2 Strength of culture is positively related to high growth. 3 Only a few values and practices were related to profitability.

(Continued)

Table 5.1 Cont'd

Reference	Organisational culture dimensions	Performance measure	Organisations involved	Respondents involved	Evidence for the culture–performance link
Gordon and DiTomaso (1992)	(a) strength of culture (b) adaptability (c) stability	Over a 6-year period: (a) growth of assets (b) growth of premiums	11 US insurance companies	850 managers	Culture strength and adaptability are both predictive of short-term performance.
Kotter and Heskett (1992)	(a) strength of culture (b) strategy-culture fit (c) adaptability	Average over 11 years of: (a) yearly increase in net income (b) yearly return on investment (c) yearly increase in stock price	207 US firms from 22 different industries	600 top managers	There is a positive but moderate relationship between culture strength and long-term economic performance.
Marcoulides and Heck (1993)	(a) organisational structure (b) organisational values (c) task organisation (d) organisational climate (e) employee attitudes	(a) gross revenue/product value ratio (b) market share (c) profit (d) return on investment	26 greatly varying US firms	392 employees	All culture dimensions have some direct or indirect effect on performance.

Study	Cultural dimensions	Performance measures	Context	Sample	Findings
Denison and Mishra (1995)	(a) involvement (b) consistency (c) adaptability (d) mission	(a) perceived performance, (b) objective performance as average over 3 years of return on assets and sales growth	764 firms in five different US industries	764 top managers	1 For large firms profitability is best predicted by stability traits such as mission and consistency. 2 Sales growth is best predicted by flexibility traits such as involvement and adaptability. 3 All cultural traits were positively related to return on assets, with mission as the strongest predictor.
Petty et al. (1995)	(a) teamwork, (b) trust and credibility, (c) performance improvement/common goals, (d) organisational functioning	(a) operations, (b) customer accounting, (c) support services, (d) employee safety and health, (e) marketing	12 service units within a US firm in the electric utility industry	832 employees	Much teamwork is associated with high performance.
Koene (1996)	(a) process vs. results orientation, (b) employee vs. job-orientation, (c) professional vs. parochial orientation, (d) open vs. closed culture, (e) tight vs. loose control, (f) normative vs. pragmatic	(a) store performance, (b) cost performance, (c) personnel performance	50 company-owned Dutch supermarket stores of a large retail chain	1,228 employees	Employee orientation and openness influence performance, both directly and indirectly, through their impact on the climate variables, general and task communication.

(Continued)

Table 5.1 Cont'd

Reference	Organisational culture dimensions	Performance measure	Organisations involved	Respondents involved	Evidence for the culture–performance link
Fey and Denison (2000, 2003)	(a) involvement; (b) consistency; (c) adaptability; (d) mission.	(a) sales growth and market share; (b) quality of products/services; (c) employee satisfaction; (d) overall performance; (e) profitability; and (f) new product development.	179 foreign-owned firms operating in Russia in six different classes of industry for questionnaire and four case studies in depth (i.e. linked studies).	179 single respondents (i.e. one employee per sample firm) drawn from top executive level.	Flexibility-orientated traits (adaptability/involvement) rated as the most important trait in determining success in Russian context.
Zhang and Liu (2003) Unpublished study still in design stage	(a) clan; (b) adhocracy; (c) market; and, (d) hierarchy.	(a) a self-developed organizational effectiveness survey instrument	5 Chinese construction companies operating in PRC (Pilot survey)	No. of respondents not yet reported but apparently drawn from corporate executive level.	Preliminary results suggest that hierarchy and clan type cultures are stronger in Chinese construction companies than market and adhocracy.
Wilderom and Van den Berg (2004)		(a) autonomy; (b) external orientation; (c) interdepartmental coordination; (d) human resource orientation; and, (e) improvement orientation.	58 Netherlands banking firms.	1,509 respondents	This exploratory study concludes that more dimensions from other studies can be added to the five to make a stronger defining set of dimensions for comparing organizational cultures.

Source: Wilderom et al. (2000: 198–199) extended by Coffey (2004).

refer to culture as 'the way we do things around here', and the Bates and Plog (1990: 7) state that 'Culture is the system of shared beliefs, values, customs, behaviours, and artifacts that the members of society use to cope with their world and with one another, and that are transmitted from generation to generation through learning'. The following specific definition of organisational culture was developed and adapted from multi-sources by Coffey (2005) and used for the specific research described later in this book:

> the informal shared values, norms and beliefs that control how individuals and groups in organisations consistently perform tasks, solve problems, resolve conflicts and interact with each other and with others outside the organization.
>
> (Coffey, 2005: 94)

As described earlier, several methodologies and instruments used previously for measuring organisational culture were examined, and the Denison Organisational Culture Survey (DOCS) was eventually selected as being the most suitable to use in the context of research within the construction industry due to its suitability for use in, and wide acceptability by, the business environment. During the preliminary discussions with representatives of the construction industry in Hong Kong as well as with consultants operating within the industry in the facilitation and partnering areas, substantial support for the use of a business-orientated model when measuring organisational culture in the construction industry context was gained, since the general consensus was that the traits and management aspects described in the model have particular relevance for the construction industry. Before describing in more detail the DOCS approach, it is necessary to examine the theories and literatures that have driven the development of the Denison model, in particular the four trait concepts of *involvement, consistency, adaptability* and *mission*.

The Denison Organisational Culture Survey (DOCS) model

The involvement concept

From the large body of literature on human relations theory, a concept emerges that the level of involvement and participation of organisational members in organisational processes and development has a direct impact on the observable degree of organisational effectiveness. Much of the organisational behaviour research undertaken in the early 1960s emphasised the hypothesis that high levels of involvement and participation created a sense of ownership and responsibility that later developed into organisational commitment, and allowed voluntary normative systems to thrive and operate within integrated, largely non-bureaucratic organisations

(McGregor, 1960; Likert, 1961; Argyris, 1964). Further research carried out in the late 1970s and in the 1980s expanded this hypothesis and posited that deliberate structuring of organisations to elicit high involvement of all staff improved performance at management and worker levels, the former due to greater strategic focus and the latter due to operating a better work environment (Walton, 1977, 1986; Lawler, 1977, 1986). Ouchi (1981) discusses high-involvement organisations in the context of their forming 'clans' where formal bureaucracies are rejected in favour of management systems where costs are minimised and efficiency is increased based on internally evolved integrated organisational structures and consensually developed operating procedures. Despite the research undertaken on the involvement–efficiency concept, more recent conclusions are that there is at best only a tenuous relationship between the participation of organisational members and measurable organisational performance (Locke and Schweiger, 1979; Miller and Monge, 1986). Denison (1990: 8) expresses more positive support and states that 'the hypothesis is a compelling one and persists as a central element in this theory of organisational culture and effectiveness'. This book examines the hypothesis further.

The consistency concept

The consistency–cultural strength–effectiveness hypothesis assumes that internally developed value-based control systems are a more effective means of achieving integration and coordination within organisations than are proscriptive, externally developed systems of rules and regulations (Weick, 1985, 1987). In the construction industry, these different perspectives might be demonstrated by examining a company that is run following the philosophies of total quality management as opposed to a company operating an ISO9000 quality system demanded of it by powerful clients for the promise of work continuity. Characteristics of companies which espouse consistency as an asset would typically be highly committed employees, strong central core values, a unique and recognisable way of doing business, the tendency to promote internally and a clear set of operational procedures with set parameters. Several authors have emphasised the importance of this concept (Deal and Kennedy, 1982, 2000; Peters and Waterman, 1982; Martin and Siehl, 1983; Frost *et al.*, 1991), but none of these studies has differentiated between the theory of involvement described earlier and the theory of consistency examined here. So what additional perspectives does the literature on consistency add to that of involvement when viewed against the importance of both to organisational effectiveness? In an early piece of research, Seashore (1954) reported that when performance norms are shared, well communicated, highly consistent and well integrated into an organisation, this is often a predictor of future organisational success. This view on the contribution to business effectiveness of strongly shared and agreed norms and expectations and their ability to control behaviour in ways that formal control systems and organisational structures are unable to do so is supported by other researchers (Moch and Seashore, 1981; Georgopoulos, 1986; Cameron and Freeman, 1989). However, Denison (1990) warns that consistency may be considered to be a 'double-edged sword', referring to the view of Pascale (1985) who concluded that strong cultures operating in an environment of norms and expectations unmatched to the business setting become a liability and may lead to inertia which is an obstruction to the organisational changes needed to drive high levels of effectiveness. A fundamental and long-held argument arising with regard to the concept of consistency in organisations driving effectiveness is that in order for this to happen organisational practices and policies must be closely aligned with the core values and beliefs of all organisational members, and all must share a desire to conform to them (Whyte, 1951).

While similarities exist between the concepts of both involvement and consistency, the main difference appears to be that involvement theory basically assumes that well-integrated organisations, which allow participation of members in the design of organisational processes, will overcome the disadvantages that sometimes result from a purely generic democratic process for decision-making. Consistency theory argues that conformity, consensus and consistency outweigh low levels of involvement and participation (Denison, 1990). According to Denison (1990) and White (1988), the most effective organisations use the cycle of the involvement of their members to develop strategies and identify solutions which are then refined into an accepted set of consistent operational principles shared by those same members.

The adaptability concept

Although not approaching the issue from a cultural perspective, several authors have examined the effects that operating environment has on the development, structure and effectiveness of organisations (Lawrence and Lorsch, 1967; Katz and Kahn, 1978; Pfeffer and Salncik, 1978; Aldrich, 1979). Starbuck (1971) and Buckley (1968) have theorised that in order to survive, organisations must adopt a complex adaptive nature and constantly restructure and alter internal processes to survive against their competition, i.e. morphogenesis. According to Kanter (1983), organisations that cannot undergo morphogenesis are unable to drive change and become heavily bureaucratic with value systems biased towards retaining stability at all costs. There are three distinct aspects of adaptability that may have an impact on the effectiveness of organisations:

1   The ability to understand their external environment and in particular external customers (Peters and Austin, 1985; Abegglen and Stalk, 1986).
2   The ability to respond to internal customers and overcome internal barriers.

3   The capacity to restructure and 'grow' a new set of behaviours to meet internal and external customers' needs and develop altered processes for new demands and environments (Zald and Ash, 1966).

Schein (1985) approaches the concept of adaptability from the cultural perspective and describes how organisations, confronted with new environments and business situations, first use their existing shared and agreed behavioural responses that have proved successful for adaptation to changed circumstances in the past. He gives a clear perception of how the socialisation processes or organisations contribute to their culture but not how an organisation's culture contributes to its ability to adapt to change. Drawing on this concept, it appears that the various aspects of adaptability are at least supported by organisational culture in terms of underpinning the contribution to effectiveness.

The mission concept

According to Denison and Mishra (1995: 216), 'relatively few authors have written directly on this topic [i.e. the importance of 'mission' to culture and effectiveness]'. They concur that there was agreement among many such authors that mission provides purpose and meaning to the functioning of organisations and also that 'possessing a sense of mission sets out a strategic route map for the organisational strategy to be followed by its members' (Selznik, 1957; Bourgeois and Eisenhardt, 1988; Robbins and Duncan, 1988; Hamel and Prahalad, 1989; Westley and Mintzberg, 1989; Westley, 1992). Locke (1968) observed that at the level of the individual, goal-directed organisational members are more likely to achieve targeted success outcomes. Weick (1979) and Davis (1987) found that a strong sense of mission requires organisations to use 'future-perfect' thinking that shapes organisational behaviour by targeting some desirable long-term future achievements for all to aim at. Fey and Denison (2000, 2003) summarise succinctly the more contemporary literature on the impact of mission on effectiveness of businesses (Ohmae, 1982; Mintzberg, 1987, 1994; Hamel and Prahalad, 1994), stating that 'successful organisations have a clear sense of purpose and direction that defines organisational goals and strategic objectives and expresses a vision of how the organisation will look in the future'. Denison (1990: 156) evolved a way to integrate these hypotheses together into a framework, and an adaptation of this simplified model is shown in Figure 5.3.

If viewed separately, each of the four concepts represent individual cultural traits that according to the relevant literature have a strong influence on organisational performance and effectiveness. Using the theories described earlier in this section (i.e. that adaptability and mission are focused on reaction with the external environment, whereas involvement and consistency represent an internal focus), the model thus forms two horizontal pairs of traits: the upper of which are externally referenced

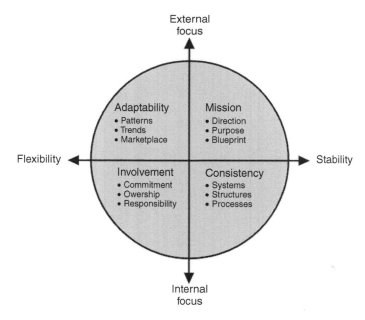

Figure 5.3 The Simplified Denison model.
Source: Denison (1990) adapted by Coffey.

and the lower internally referenced. The vertical pairing of adaptability and involvement is related to the capacity of an organisation for change and flexibility and the other vertical pairing of mission and consistency reflect the stability and strategic direction of the organisation. Denison (1990) acknowledges that there is some similarity between his model and the Competing Values Framework Model of his contemporaries (Quinn and Cameron, 1988) and justifies this by noting that 'the reconciliation of conflicting demands is the essence of an effective organisational culture' (p. 15) and also that 'as Quinn has noted...it is the balancing of competing demands that distinguishes excellent managers and organisations from their more mediocre counterparts' (p. 16).

Based on DOCS data obtained from over 1,000 high- and low-performing organisations following the first development of the simplified model described above, Denison and Neale (1994) subsequently found that the four culture traits were further subdivided into 12 'management practices' as follows:

- *Involvement*

    - Empowerment
    - Team orientation
    - Capability development

- *Consistency*
  - Coordination and integration
  - Agreement
  - Core values

- *Adaptability*
  - Creating change
  - Customer focus
  - Organisational learning

- *Mission*
  - Strategic direction and intent
  - Goals and objectives
  - Vision.

Figure 5.4 shows all the operational dimensions of the modern DOCS resulting from subsequent work by Denison and his colleagues (Denison and Mishra, 1995; Denison and Neale, 1996; Denison *et al.* 2000, 2004; Fey and Denison, 2000, 2003).

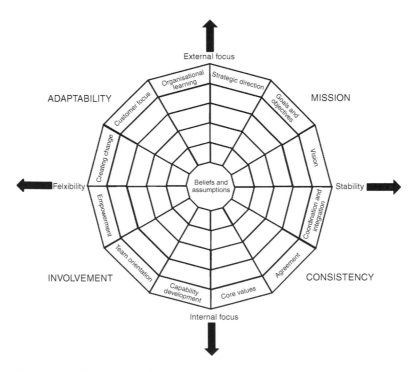

Figure 5.4 The Denison Organisational Culture model.
Source: Denison (2009) adapted by Coffey.

A subsequent paper by Denison *et al.* (2000) presents a validity test of the 60-item, 12-index DOCS which was developed to measure the model's key constructs using responses from a sizeable population (36,542 individuals in 94 organisations). The results from confirmatory factor analysis tests and evaluation of structural equation models supported both the measurement model itself and the theoretical structure implied by the framework. The operationalisation of strong cultural scores used in this book is based on companies that had individual or combined OC trait scores on the third and fourth quartile of the DOCS. The concept of organisational effectiveness used in this book is different to that used in any of the previous research studies referred to in this chapter. This is mainly due to the criticisms evident in the literature concerning the previous predominant use of financial measures to evaluate company success and thus organisational effectiveness, coupled with the fact that companies being evaluated were in the construction sector and there was an apparent lack of available financial information specifically related to their work in the public housing sector (i.e. most of the companies were not publicly listed and had parent companies whose financial results were generated based on their *total* development sector performance). In an attempt to utilise measures which were determined on a non-financial basis and also reflected 'customer satisfaction', the objective *dependent variables* used to operationalise 'organisational effectiveness' in this research were the success ratings of building contractors employed on public sector housing contracts awarded and operated by the Hong Kong Housing Authority (HKHA). These variables were measured by means of the Hong Kong Housing Department (HKHD) Performance Assessment Scoring System (PASS) 1997 Edition, which measures objectively the quality of constructed output against the specification requirements. The specific use of the 'Denison Organisational Culture Survey' (DOCS) and the HKHD's PASS (1997) within the context of research in the Hong Kong public housing sector of the construction industry are described in more detail in Chapters 6 to 9.

Strategic performance measures in the construction industry

Construction performance measures

Much of the recent research in the construction area has centred on examining the benefits of partnering; more recently however, another huge research effort has focused on investigating the efficacy of various new paradigms being used in the construction field to drive and develop performance improvements. The development of new, and enhancement of existing, initiatives such as partnering, lean construction, TQM, JIT, concurrent construction/engineering, design and build and others has emanated from the desire to lift the image of an industry, which, according to

Forbes (2001: 54), alluding specifically to the US, 'has yet to display unanimity in its approach to quality and continuous improvement … [where] in the majority of construction projects across the country, wasted materials, wasted motion and wasted human resources are the norms'. This rather negative view of the general standard of construction industry performance has prevailed globally and has led to national governments developing their own strategic programmes and initiatives to combat the unacceptable spectres of rogue builders, shoddy workmanship and dissatisfied customers. In the executive summary of Sir John Egan's ground-breaking 1998 report *Rethinking Construction*, the following statement was made:

> The UK construction industry at its best is excellent. Its capability to deliver the most difficult and innovative projects matches that of any other construction industry in the world … nonetheless, there is deep concern that the industry as a whole is under-achieving. It has low profitability and invests too little in capital, research and development and training. Too many of the industry's clients are dissatisfied with its overall performance.
>
> (DTI, 1998: 4)

The European Construction Institute's website summarising Sir Michael Latham's April 2003 *Construction in Europe* keynote address states,

> There is no room for complacency. Construction could do better. European construction is under pressure. Growth has been sluggish for several years in a number of the biggest countries. The industry must address some key issues … the cost of non-quality in construction at 5–10 per cent of the industry's total output – that is, €40 to 80bn each year is wasted … We know that good practices work, but their uptake is still too limited … construction still has a poor image and finds it difficult to recruit and retain people of sufficient calibre, particularly for site operations.
>
> (ECI webpage, 2003)

In Hong Kong, a 2001 report entitled *Construct for Excellence* and produced by the specially convened Construction Industry Review Committee echoed similar sentiments:

> The construction industry has over the years produced numerous examples of outstanding architecture and engineering excellence. It has collectively contributed to the remarkable social and economic transformation of our society. But there is room for improvement in its overall performance in terms of quality, efficiency, productivity,

site safety, environmental sustainability and customer satisfaction. The recent spate of non-compliant construction incidents has prompted widespread public concern about the need for reforming the industry ... to the general public, construction is frequently associated with poor workmanship, budget over-runs and programme delays.

(CIRC, 2001: 7)

This same report also mentions the word 'culture' 41 times, either in the context of its needing to be changed, improved or reviewed in the Hong Kong construction arena (CIRC, 2001). In the UK, the Department of Transport and Industry has over the past decade developed various schemes to facilitate clients and contractors to improve the quality of constructed projects. Some of the major schemes may be observed from the list below; many have appeared only in the past two or three years. The reports that triggered such initiatives were:

- *Constructing the Team: Final Report of the Government / Industry Review of Procurement and Contractual Arrangements in The UK Construction Industry* (HMSO, 1994)
- *Rethinking Construction* (Egan Report, 1998)
- *Accelerating Change: A Report by the Strategic Forum for Construction* (DTI, 2002)
- *Quality Mark Scheme Member's Guide* – the register of approved builders and tradesmen working in the domestic repair, maintenance and improvement sector (DTI, 2002)
- *Partners in Innovation Scheme* – an annual competition to secure funds for construction industry research (DTI, 2003)
- *Construction – A Sector Study* (CE – Construction Best Practice, 2003)
- *Movement for Innovation* (M4I, 2000)
- *Construction Industry Key Performance Indicators* (DETR, 2000)
- *Housing Quality Indicators* (BRE, 2000)
- *Respect for People: A Framework for Action* (Rethinking Construction Ltd, 2002)

Much of the research carried out into the human relational/organisational facets of the construction industry in recent years has been concerned not only with examining team working, but has more specifically investigated the benefits of, and ways to improve the outcomes of, partnering. 'Partnering', according to the UK Department of Trade and Industry's website (DTI, 2003) 'is a label used for a variety of innovative approaches to managing relationships between organisations in construction'. Although this book is not focused on an examination of partnering, it is noteworthy that in the construction industry, described by Cheng *et al.* (2000) as 'an adversarial business', Bresnen and Marshall (2000) state their belief that partnering can significantly improve construction project performance, and this supports the work of Sommerville *et al* (1999: 729), who state that in an industry that

> has long been seen as ... highly fragmented and subject to significant cyclical changes ... new team working and best practice models will appear rapidly once the prevailing culture of the industry is better understood. [Of particular significance to my book is their support for identifying and utilizing] 'specific business excellence measures ... [and] ... metrics which will prove accurate in measuring and portraying the underlying cultural paradigm.'

Many of the reports on construction innovation and partnering/team building both in the UK and elsewhere in the world highlighted a common theme of 'the need to change the culture of the construction industry', as shown in Table 5.2.

The need for cultural change in the construction industry

It seems somewhat incongruous therefore that, given the rather public declarations of this need for culture change in the construction industry, relatively little research has so far been done to try to establish and measure what that culture actually is and how it relates to better or worse performance of construction companies. In the Preface to an International Council for Research and Innovation in Building and Construction (CIB) report entitled *Perspectives on Culture in Construction* this 'critical gap' in the area of culture studies was summed up by Fellows and Seymour (2002), who state that:

> The need for formalizing research into the role and impact of culture within the boundaries of the construction industry ... became clear at the AGM of the Association of Researchers in Construction Management (ARCOM) .... held in the Isle of Man in 1993 ... a particular aspect of the work of TG-23 [the CIB task group specifically charged with investigating *Culture in Construction* and now CIB Working Group W112] which has become increasingly clear ... is that the applicability of this research area [i.e. construction culture] is not only generic, but more importantly, is a critical arena in its own right.
>
> (Fellows and Seymour, 2002: i)

Research into organisational culture in the construction industry

As mentioned above, CIB's former TG-23 has certainly been a global main force in launching and supporting construction-specific studies of

Table 5.2 Recent reports referring to the need for culture change in the construction industry

---

'If the industry is to achieve its full potential, substantial changes in its *culture* and structure are also required to support improvement. The industry must provide decent and safe working conditions and improve management and supervisory skills at all levels. The industry must design projects for ease of construction making maximum use of standard components and processes' *(Rethinking Construction, 1998, p. 5)*.

'The aim of this programme is to develop a culture for change and improvement. This is essential if we are to reap the full potential of the technical and process innovations now being implemented across the construction industry .... Further, industry needs to be persuaded that people and *culture* issues are worthy of investment .... Key areas for research could include:

–   compare *culture* and people management practices in construction with those in other industries' *(Partners in Innovation Programme – Call for Proposals and Guidelines for Applicants, 2002, p. 2)*.

'To achieve a step improvement in its overall performance, the construction industry needs to develop a new *culture* that focuses on delivering better value to the customers on a continuous basis. We have a vision of an integrated construction industry that is capable of continuous improvement towards excellence in a market-driven environment' *Construct for Excellence (Construction Industry Review Committee, 2001, p. 8)*.

Continuous improvement *culture* [means that] .... To become truly innovative the industry must commit to continuous improvement in production and deliver innovative solutions to meet clients' needs. It must build a workforce, which is flexible, skilled, continuously educated and mobile. The industry must work smarter by re-engineering business and project processes and by engaging in research and development' *(Construct New South Wales, 1998, p. 28)*.

'The building and construction industry in Australia requires significant cultural change. Such change is necessary if the rule of law is to be reintroduced to conduct and activities within the industry, if individuals' freedoms are to be maintained, and if the industry is to achieve its economic potential. Change is required in the attitudes of all sectors of the industry, including governments, clients, head or subcontractors, industrial organisations and employees' *(Final Report of the Royal Commission into the Building and Construction Industry: Vol. 11 – Achieving Cultural Change, 2003)*.

'Lasting change in an industry, workplace, organization or society is only possible if the culture from which the behaviour emanates itself changes .... All of this reinforces the point that deep seated unacceptable cultures may take many years to change and that they involve a combination of education and strong regulation and sanctions' *(Inquiry into Buildings and Construction Industry (Restoring Workplace Rights) Bill, 2008)*.

'The trend towards consideration of non-price criteria and the advent of relationship management and alliance-type contracts has encouraged increased focus on the collaborative elements of project team management. Industry accepts that a cultural shift is required to maximise the outcomes from such projects. However, fostering the right culture is not a challenge for the project team alone. The client organisation must also develop an appropriate culture to be able to propose and manage relationship contracts' (Relationship management in the construction sector – culture change) (CRC Construction Innovation website – 2006).

Sources: As cited, collated by Coffey (2005).

organisational culture and performance, and over the past decade some culturally based studies pertaining specifically to the construction industry have begun to emerge. As the number of research articles and publications is still relatively small compared to the general literature pertaining to cultural and organisational studies, it is practical to examine the major ones in the following section. According to Root (1994: 1):

> in the study of the construction environment, very little attention has been paid to the question of culture as an environmental factor or influence [and perhaps more importantly within the context of my own book, he acknowledges the fact that management of construction projects at national levels requires an awareness and understanding of the] ... values and priorities, assumptions and attitudes, expectations and habits of mind that are developed within different occupational and corporate groupings.
>
> (Root, 1994: 1)

The 1991 Engineering and Physical Sciences Research Council (EPSRC) funded research project led by Root was further developed into a Ph.D. (2001), which as one of its fundamental research questions enquired whether such cultural values and attitudes among project teams had a significant effect on how the project proceeds (Root, 2001). However, this differs from the research by Newcombe *et al.* (1990) where, using their own systems model developed for the construction production process, culture was shown not to be a significant environmental factor influencing the process. A contemporary study by Bresnen and Marshall (2000) into the effects of partnering within the construction industry upon project performance also studied the complex organisational relationships present within construction companies and how these impacted on their organisational culture. The authors advocate the need to develop not only 'specific business excellence measures' within the industry, but also a set of measures to establish 'key culture parameters' of construction organisations in order to develop models to integrate business excellence with culture change and business strategies. They posit that the application of a 'polar plot model' to ascertain generic team culture plots and subsequent benchmarking against other successful project teams and organisations in the field will enable companies to undertake more successful partnering on projects, share organisational learning and still remain competitive. Work is also being undertaken to critically examine the procurement systems operated by clients to select its construction project partners (Kumaraswamy and Dissanayaka, 1996; Holt and Proverbs, 2001; Wong and Holt, 2001). Lansley (1994) in a definitive paper examined in depth various methodologies for 'analysing' construction organisations and states that when applied to construction organisations the extensive literature on organisation theory can be confusing and conflicting (Lansley, 1994: 337). The paper describes in depth the various studies undertaken to explain how construction organisations function, and the author suggests (Lansley, 1994: 343) that although these studies have used quite different perspectives and vary in their specific conclusions, there does appear to exist evidence of a link between the better established theories and their associated models of organisation. Seymour and Rooke (1995) have examined in depth the state of both the organisational culture of the construction industry and of research into the concept of culture within that industry. Maloney and Federle (1991) utilised the Competing Values Framework model to assess and analyse the culture prevalent at the management levels of US engineering companies, and Gale (1992) has examined the question of the effects of 'deculturalising' (i.e. the replacement of male cultural characteristics and stereotyped attitudes with feminine thinking processes) of male construction operatives, as an intervention technique in conflict resolution on building projects. Several researchers have begun to focus on both the impact of cultural differences and of the global differences in cultures, as they appear in international construction projects. Lingard and Rowlinson (1998) and Rowlinson and Lingard (1998) have written extensively about the establishment and operation of safety culture in the construction industry, particularly in Australia and Hong Kong. Hall and Jaggar (1997) and Abeysekera (2003) examined the effects of national cultural differences in international construction projects. This was also the focus of a study of the Swedish construction industry by Bröchner, *et al.* (2002) who concluded that many of the dimensions first discovered by Hofstede (1991) were present and of significance, and that any improvement in the quality outcomes of construction projects in Sweden would require a major cultural shift towards greater trust in partnering techniques and less reliance on non-cooperative contract relationships. Liu (1999) investigated the relationship between cultural traits and job satisfaction in the real estate profession in Hong Kong and noted that 'organisational culture appears to be gaining support as a predictive and explanatory construct in organisational science'. Liu supports the proposition that, once identified, organisational culture dimensions (in her paper, within the real estate profession) can be shaped to impact positively on the job satisfaction of individuals. Using a nine-dimensional representation of organisational culture, Liu (1999) verified that 'team-oriented ... communicative/supportive cultures have a positive role in enhancing real-estate professionals' job satisfaction'. This was contrary to an earlier study by Noyes (1992), which concluded that the culture of the real estate industry was somewhat weak due to the finding that estate agents appeared lacking in any strong sense of integration within teams, i.e. no group identity. Rowlinson (2001) undertook an in-depth study of the structure (and culture) of a major Hong Kong Government public sector development body and described its change to a matrix-based organisation, and Rowlinson and Root (1996), in an as yet unpublished paper, investigated the impact of culture on project management outcomes and success factors.

Liu (1999) and Liu and Fellows (1999a, 1999b) have authored three articles, one of which examines the impact of culture on project goals, another on the impact of culture in the Hong Kong real estate profession and the third examines the generic and specific issues of culture related to the construction project procurement systems.

In the past few years, some of the studies summarised here have been further extended and some new culturally related issues have also appeared in the construction research arena. Barthforth *et al.* (1999) reviewed the extant literature of both construction industry and organisational culture and performance studies, and developed new meanings and defined three different concepts of culture, specifically related to the construction industry, and exposed issues of concern within the industry using the three concepts as a framework. Liu (2002) has explored the Eastern cultural trait of 'harmony' as it pertains to project management and is critical of the lack of clear theoretical constructs for examining the phenomenon in the construction literature. Liu (2002: 35) posits that the concept of 'harmony' is an important attribute to successful project partnering and advocates the use of triangulated methodologies including a more ethnographic (i.e. cultural) approach to future studies of such phenomena in construction scenarios.

Chapter summary

This chapter has described and investigated the development of research into the link between organisational culture and effectiveness over different periods in the history of this thematic development. The Denison Organisational Culture Survey was introduced in detail and the chapter ended with a short review of those few studies into culture in the construction industry that have been published in recent years.

6 Research on the relationship between organisational culture and performance in Hong Kong construction companies

Chapter introduction

An introduction to current research on organisational culture within the construction industry concluded the last chapter providing a platform from which to move to the material in this chapter and those that follow. Earlier chapters of this book have taken a detailed journey through the origins of organisational research, the concept of organisational culture and organisational effectiveness and the linkage between these concepts, as well as the importance of such linkage. Such a journey is necessary for the reader to be fully prepared to 'understand' what now follows in this and in subsequent chapters – a description of research carried out among construction companies involved in one of the largest public housing delivery programmes in the world and an analysis of what the findings of the research mean for the construction industry at large. This chapter is relatively long in order to keep the research elements covered in this book together, but in order to make the text easier on the reader; the chapter is split into three main sections as follows:

- Part I – Research questions
- Part II – Research instrument
- Part III – Data analysis methodology/administration.

PART I

Research questions

Emanating from the literature review described in earlier chapters of the book and based on the author's intimate working knowledge of the prevailing population of Hong Kong contractors engaged in public housing construction, three main research questions developed, which translate into nine hypotheses.

In order to better illustrate both, two simple models have been developed, the first to represent the research questions and the second the research hypotheses. The theoretical framework of the research questions already

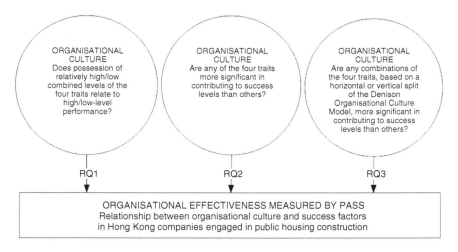

Figure 6.1 Theoretical framework of research questions on the link between organisational culture and effectiveness.

presented in Chapter 1 is shown again here in Figure 6.1, as it integrates closely with Figure 6.2.

The research model operationalising the nine hypotheses that were developed from the three research questions is shown in Figure 6.2.

The three research questions and nine hypotheses are:

Question 1 and associated hypothesis

Do Hong Kong construction companies possessing relatively high combined levels of the four organisational cultural 'traits' (i.e. adaptability, involvement, consistency and mission) (as indicated by the Denison Organisational Culture Model (DOCM)) perform more successfully on public housing projects than those exhibiting lower levels of those traits?

H_1: Position of 'successful' building contractors on the PASS score league is significantly associated with high (i.e. third and fourth quartile) combined organisational culture scores, as measured by the DOCM.

Question 2 and associated hypotheses

Are any of the four traits more significant in contributing to success levels than others?

H_2: Position of 'successful' building contractors on the PASS score league, is significantly associated with 'adaptability' in their organisational culture as measured by the DOCM.

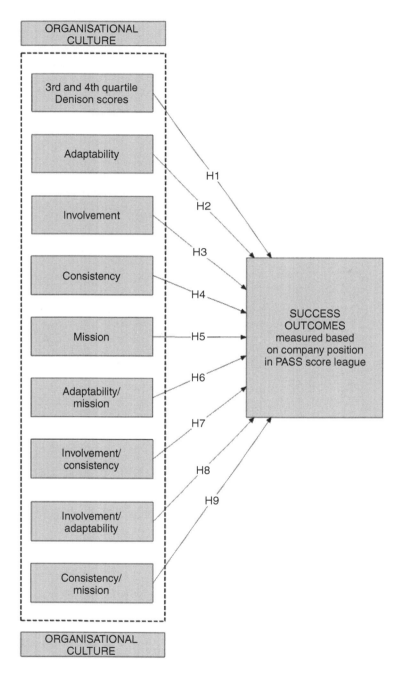

Figure 6.2 Research model operationalising the nine hypotheses.

H₃: Position of 'successful' building contractors on the PASS score league is significantly associated with 'involvement' in their organisational culture as measured by the DOCM.

H₄: Position of 'successful' building contractors on the PASS score league is significantly associated with the 'consistency' measure in their organisational culture as measured by the DOCM.

H₅: Position of 'successful' building contractors on the PASS score league is significantly associated with the perception of 'mission' in their organisational culture as measured by the DOCM.

Question 3 and associated hypotheses

Are any combinations of the four traits, based on a horizontal or vertical split of the DOCM, more significant in contributing to success levels than others?

H₆: Position of 'successful' building contractors on the PASS score league is significantly associated with third and fourth quartile 'Denison' rankings in 'adaptability/mission'.

H₇: Position of 'successful' building contractors on the PASS score league is significantly associated with third and fourth quartile DOCM rankings in 'involvement/consistency'.

H₈: Position of 'successful' building contractors on the PASS score league is significantly associated with third and fourth quartile DOCM rankings in 'involvement/adaptability'.

H₉: Position of 'successful' building contractors on the PASS score league is significantly associated with third and fourth quartile DOCM rankings in 'consistency/mission'.

Research methodology

This section describes the major and secondary methodologies used to collect data and introduces the methodology for the subsequent analysis of the data to investigate and answer the research questions raised earlier in the chapter.

Research paradigm

The major paradigm of the research is functionalist and the viewpoint objectivist, because the primary data obtained from the Denison Organisational Culture Survey (DOCS) were based on an instrument that uses quantitative rating scales and from the secondary data drawn from the rankings and scores obtained from the Hong Kong Housing Department (HKHD)'s PASS, an objective and quantitative measurement tool. Thus the worldview for this research is realist/positivist, nomothetic and, to a degree, deterministic. It follows therefore that the overriding research methodology also locates into the positivist (i.e. the quantitative) area. The mainstays of the methodology involve the use of a well-validated survey instrument, adapted for the particular location and environment in which it was to be used, which collected perceptions and measured the strength of four organisational culture traits in the sample population. The results from the DOCS were then correlated with performance factors drawn from specific PASS project data representing the effectiveness of companies forming the original survey sample after these factors had undergone discriminatory factor analysis to ascertain their actual efficacy in measuring 'success'.

Using this information, the research seeks rational explanations for the actions of members of the organisations in the sample population (Burrell and Morgan, 1992). At this point of the research it is useful to consider its philosophical perspective, which while based mainly on findings from analysis of quantitative data, also relies to an extent on qualitative findings and therefore sits astride two major research paradigms. The following descriptions are derived from Peter Rouncefield's website *Sociology Stuff* (2004) and provide a concise description of the positivist and interpretivist paradigms, and their particular relevance to this research is briefly discussed.

(Logical) positivism, as a research methodology, asserts:

- the pre-eminence of direct, systematic observation;
- that knowledge should be based on the 'positive' facts of experience;
- that causal explanation should be arrived at by way of inductive generalisation;
- scientific knowledge is founded on facts (i.e. positivism is a scientific method);
- that knowledge is not formed just from normative statements;
- a rigid objective distinction between facts and values.

As regards the study of society, positivism argues:

- that society and its constituents are an objective reality with a structure and functions that obey a set of 'rules';
- that individuals view this objective reality as a set of 'inescapable necessities';
- the research methodologies of natural science are directly applicable to the social world because:
  - society as a whole possesses measurable social facts;
  - social life can be subject to causal explanation.
- By applying investigative methodologies that have proved successful in the study of nature, namely observation, experimentation and comparison to social research, it is possible to make predictions about aspects of society.

Rouncefield (2004) makes clear that in his view the objectivity described above is in the methodology of carrying out research, and notes that most researchers follow their own personal interest in selecting what to research, or the interest of the organisation paying for the research. However, as pointed out by Sekaran (1992), 100 per cent of scientific research is not always viable in the management and behavioural areas, but the author asserts that the application of good, rigorous, objective research design that can be tested, generalised and replicated constitutes an approach that is as near to scientific research as most investigators can endeavour to attain.

The view of the interpretivist paradigm is diametrically opposed and posits that traditional scientific methods are not applicable for use in investigating human social behaviour. The literature argues that the natural and the human sciences are based on a different set of logical tenets and that the idea of a 'science of society' is untenable (Rouncefield, 2004).

Max Weber was probably the most famous proponent of this latter view and so Coser (1977), writing about Weber, states that he (Weber) vehemently rejected the positivist perspective (i.e. that social and natural sciences were the same) and posited that man (as opposed to 'things') could be understood not only through his behaviour but also by his underlying motivations. He highlighted in his writing the value-bound problems chosen by researchers and the value-neutral methods of social research, and he observed that the differences between natural and social science research arise because of the differences in the cognitive intentions of the researcher and not because of any inapplicability of scientific investigation of human actions. Rouncefield (2004) describes the assertions of interpretivists (or constructivists) as follows.

The methods used to study the natural sciences are inappropriate to the study of social life because:

*   significant sociological behaviour has meaning;
*   humans make choices because they are active and conscious beings;
*   real understanding has to involve reasoning and cannot just be based on descriptions;
*   society actively interprets the world around it rather than just simply responding to external stimuli;
*   data does not 'speak for itself', it requires interpretation;
*   societal and human behaviour is intentional and not accidental.

Rouncefield (2004) also describes constructivism as 'an educational philosophy falling within the rationalist philosophy' and summarises its following key facets:

*   reality is constructed;
*   knowledge is dependent on the knower's frame of reference;

- knowledge consists of only ideas or representations about reality;
- roots of constructivism can be traced back to Jean Piaget.

Constructivism is often further broken down into individual and collective perspectives (Brooks and Brooks, 1993):

Individual constructivism

- knowledge is constructed from experience;
- learning results from a personal interpretation of knowledge;
- learning is an active process in which meaning is developed on the basis of experience.

Social constructivism

- learning is collaborative with meaning negotiated from multiple perspectives.

Contextualism

- learning should occur in a realistic setting;
- testing should be integrated into the task, not a separate activity.

However, while undertaking the review of the literature on research method-ologies and in particular on undertaking organisational cultural research, it became evident that the debate concerning the merits of quantitative vis-à-vis qualitative methodologies was very relevant to considering how the research described in this chapter was to be conducted.

So if the positivist paradigm utilises quantitative methods and the constructivist paradigm underlines qualitative methods, according to Tashakkori and Teddlie (1998) the use of both in a 'mixed' model is known as pragmatism, and these authors state that this mixed model combines both the positivist and constructivist approaches. Denzin (1989) states that mixed methods offer the 'best of both worlds' in drawing upon the strengths and offsetting many of the limitations of quantitative and qualitative methods. However, the methodology has some critics and this emanates from the debates that arose from the paper by Burrell and Morgan (1992), who discuss the incommensurability of paradigms (i.e. that research based on different paradigms is thus also based on different ontologies). Deetz (1996) has postulated that Burrell and Morgan's work has caused an unproductive focus on paradigm issues rather on the nature and purpose of researching particular phenomena. Skandura and Williams (2000) have noted a sharp increase in research strategies that compromise both triangulation and validity and so, while not without its critics (Cicourel, 1974; Silverman, 1985; Lincoln and Guba, 1985; Denzin, 1989; Patton, 1990), it appears that many other authors support the use of a 'multi-methodological' approach where a single study combines both quantitative and qualitative methods and uses primary and secondary data (Kaplan and Duchon, 1988; Lee, 1991; Gable, 1994; Denzin and Lincoln, 1994, 1998; Deetz, 1996; Saunders *et al.* 2000). Many recent organisational studies also strongly support and validate the method (Brown, 1998; Benjamin, Greene and Casacelli, 2003; Petter and Gallivan, 2004), and it is my belief that the major advantages of employing multi-methods in the same study are that each method may be used for different purposes and it also enables methodological triangulation to take place. This leads to personal strong support for the view of Wilder (2002: 60), who states that 'Replication is the best mechanism for validating the 'confirmability' of a research effort ... as reliability and validity are ingrained in a research project, confirmability becomes a necessary metric for scholarly acceptance.' Triangulation refers to the use of different data collection methods within one study in order to ensure that the data are actually telling you what you believe they are telling you and therefore it is not an 'adjustment' of a positivist view; it is central to the post-positivist philosophical paradigm, or as noted by Trochim (2001: 19),

> Because all measurement is fallible, the post-positivist emphasizes the importance of multiple measures and observations, each of which may possess different types of error, and the need to use *triangulation* across these multiple error sources to try to get a better bead on what's happening in reality.

According to Denzin and Lincoln (1998: 3),

> Qualitative research is multi-method in focus, involving an interpretative, naturalistic approach to its subject matter. This means that qualitative researchers study things in their natural settings, attempting to make sense of, or interpret, phenomena in terms of the meanings people bring to them.

In their 1994 book on qualitative methods and analysis, Symon and Cassell set out a number of defining characteristics of qualitative research, including '... a focus on interpretation rather than quantification, an emphasis on subjectivity rather than objectivity' (Symon and Cassell, 1998: 7). In his review of their book, Dachler (1997: 710) asserts, 'true qualitative research is based on subjectivist ontology and a constructivist epistemology'. Trochim (2001:20) also supports this view when he states,

> Most post-positivists are constructivists who believe that you construct your view of the world based on your perceptions of it. ... The best hope for achieving objectivity is to triangulate across multiple fallible

perspectives. Thus objectivity is not the characteristic of an individual; it is inherently a social phenomenon.

One of the most compelling discussions supporting the multi-method approach is found in Easterby-Smith *et al.* (2002: 31), who state that 'increasingly authors and researchers who work in organisations and with managers argue that one should attempt to mix methods to some extent, because it provides more perspectives on the phenomena being studied', and the same position is recommended by Gable (1994). Sackmann (1991) provides a summary and discussion on the pros and cons of different research methods used to investigate and analyse organisational culture and then establishes a continuum for the different methodologies. According to De Witte and van Muijen (1999), the range of methods for studying culture extends from pure ethnographic studies through participant observation, in-depth interviews with individuals as well as with groups to highly structured approaches using questionnaires and structured interviews, and the extremes of the range represent the two opposite perspectives of the 'insider' and the 'outsider'. These authors then go on to discuss the merits and demerits of both perspectives, eventually opting for a complementary approach that combines the strengths of both methodologies. They conclude that a combination of a quantitative instrument such as a questionnaire or survey followed up by dissemination of the results to members of the organisation and then following that up by interviewing and involving them in further group discussion best combines the insider and outsider perspectives, giving richer insights into the organisation's culture. Even in the contemporary social marketing literature, strong support for integration of methods may be found, according to Weinreich (1996: 53):

> The greatest weakness of the quantitative approach is that it decontextualizes human behavior in a way that removes the event from its real world setting and ignores the effects of variables that have not been included in the model. Qualitative research methodologies are designed to provide the researcher with the perspective of target audience members through immersion in a culture or situation and direct interaction with the people under study.

The view expressed by Denison and Mishra in the final paragraph of their 1995 paper that examines the relationship between organisational culture and effectiveness is one of the strongest supportive arguments encountered with direct relevance to the research described in this book. The authors state:

> Finally, in our attempt to identify some core traits in a theory of organisation culture and effectiveness, we have generally paid only limited attention to the broader cultural contexts within which the

organisations themselves exist. Societies, industries, occupations, regulatory environments, to name only a few, generate cultural contexts that influence organisations and their effectiveness. Further progress toward a general theory of organisational culture and effectiveness will clearly require that these factors will be incorporated.

(Denison and Mishra, 1995: 221)

Greene *et al.* (1989) identified 57 studies that employed both quantitative and qualitative approaches, and Brewer and Hunter (1989) noted that multiple methods were being used in most major areas of social and behavioural science research, an observation supported by Tashakkori and Teddlie (1998: 128), who refer to triangulation as 'parallel analysis of QUAL [qualitative] and QUAN [quantitative] data'. They go on to note that in educational research it is usually the norm to collect structured survey or other quantitative data (e.g. formal measures of classroom behaviour) and concurrently collect more qualitative information (informal school observations or interviews). Other authors in the education and evaluation research fields (Howe, 2003; Reichardt and Rallis, 1994) refer to the 'pragmatic paradigm' and have adopted an approach referred to as 'paradigm relativism', i.e. using whatever philosophical/methodological approach suits the research. Various research designs are discussed in the literature that combine or mix methodologies and many of these are particularly pertinent to the study of more complex organisational attributes such as culture. For example, focus group interviews are seen to be a valuable way of triangulating data collected by what are generally considered as quantitative means, (e.g. questionnaires or structured surveys) and, more importantly, allowing participants to share their perspectives of values and beliefs (Morgan and Krueger, 1993; Steyaert and Bouwen, 1994). The following salient comments are made on the use of the mixed method approach in relation to a current project investigating the internal organisational culture at Kent State University, Ohio, carried out by an HR consultant Cultural Research, Inc. (2004):

> As with studies that primarily are focused on the quantitative or qualitative perspective, for an organisational culture study the focus of validity is concerned with whether the concepts under study are of concern to the 'real world.' Each paradigm has proven methods of probing the 'real world.' For a cultural study to be considered valid, the participants in the study must agree that the 'real world' is as they reported it, and as described and interpreted by the research team.

Based on these various discussions and in view of the fact that the research described here is essentially 'an organisational culture study', the multi-methodological approach has been used and the quantitative primary survey and secondary effectiveness data has subsequently been supplemented and triangulated with qualitatively based mini-case studies, undertaken in four construction companies engaged in the production of public housing in Hong Kong. Three of these four companies were also ranked at the top end of the Hong Kong Housing Department (HKHD)'s Premier League System.[1]

However, in order to more rigorously investigate the relationship between stronger and weaker organisational cultures and effectiveness, a further focus group interview session was held in the worst-performing company of the four companies examined in the mini-case studies. This was done primarily to validate the range of the dimensions already identified in the relationship hypotheses, which existed in the sample population and also to ascertain if the antithesis of the original research hypothesis is true, i.e. do companies which do not perform well have weak organisational cultures? Relating back to the previous review of the literature, in earlier chapters of this book, this was an important part of the research process, as it has been suggested (Denison, 1990, 2000; Denison and Mishra, 1995) that much culture performance research has looked only at strong performers and ignored the weaker companies.

The research population

The research described here was carried out on 23 of the 53 construction companies listed in the Hong Kong Housing Authority's List of Building Contractors (Building – New Works Category). The main target groups within the research population were senior executives, middle managers and contract managers who are direct employees of the companies and have been educated to diploma/certificate/Bachelor's or Master's degree levels or above. According to the demographic results retrieved from the Premier League System (PLS) exercise which was carried out prior to the time of the survey, this target group represented around 50 per cent of the 4,289 total directly employed staff of all 27 PLS contractors. Based on the analysis of the returned questionnaires used as the basis of data collection for this thesis, as will be seen later in this chapter, around 87 per cent of the respondents fitted the desired managerial level and educational profile. These particular members of staff were targeted within companies in an attempt to eliminate possible 'noise' factors that could result if the study reached down to *indirectly* engaged multi-layers of domestic subcontractors. Based on the results from the PASS Management Input raw scores, together with the recorded observations of project staff of the HKHD, from the 27 companies over a period of years, the relationship between the main contractors' and subcontractors' organisations is generally not close either in management, strategy or shared corporate value terms. This set of relationships may be worthy of further study, a point that is expounded in the final chapter.

Multi-layered subcontracting – significance to the research

The phenomenon of multi-layered subcontracting is not only restricted to Hong Kong; it is prevalent in South East Asia generally, when compared to say, the UK, Europe or the USA (Lambert *et al.* 1996; Wong, 2003; Battersby, 2004). The Singapore Ministry of Manpower and Ministry of National Development C21 report (1999), *Re-inventing Construction*, identified the main problems of Singapore's construction industry. It blamed the industry's low productivity level, negative productivity growth and poor safety on the heavy reliance on unskilled foreign workers and labour intensity of construction works, and multi-layered subcontracting. In a 1998 survey by the Japanese Institute for Construction Economics the authors state that 'many criticisms have been made of the multi-layer structure of subcontracting, over half the construction operators who responded saw problems with the system. 22.3 per cent said the system was necessary but inefficient, while 33.8 per cent wanted to abolish the system' altogether. A Hong Kong consultative paper, *Quality Housing: Partnering for Change*, while recognising the need for subcontracting (HKHA, 2000: section 7.6) stated, 'Main contractors have to engage specialist sub-contractors to assist them in the delivery of projects. Sub-contracting also provides a flexible means of meeting fluctuating workload and maximizing the utilisation of expertise in the industry.' The section went on to make the following observations, stating that:

> While recognizing the need for the sub-contracting system, we notice that some unscrupulous main contractors and sub-contractors have assigned their entire jobs to others. They act as little more than brokers. This kind of unrestrained multi-layered sub-contracting activity has given rise to two main problems –
>
> (a)   Main contractors have lost control over the quality and progress of works by sub-contractors.
> (b)   Because of the profiteering activities in-between different contractors, the final delivery agents have to work on unrealistically low budgets and are hence tempted to cut corners.
>
> (HKHA, 2000: section 7.7)

The report gave an indicative view of around three layers as constituting a typical subcontracting arrangement; however, from the author's experience, the reality is that such arrangements extend to at least another three to six levels. Figure 6.3 illustrates the arrangement as outlined in *Quality Housing: Partnering for Change* (2000).

Indeed, one of the HK construction industry's newer technical forums, the Provisional Construction Industry Co-ordination Board, in its inaugural 2002 report states that:

> Subcontracting is a long-standing practice in the local construction industry that provides an essential element of flexibility in the overall

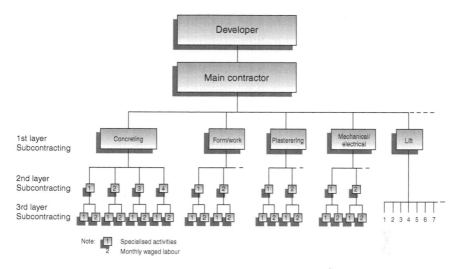

Figure 6.3 Multi-layered subcontracting arrangements in the Hong Kong construction industry.
Source: HKHA (2000)

supply chain. However, it has been plagued by a host of problems, notably those associated with multi-layer subcontracting which blurs the accountability for assuring built quality and non-productive subcontracting which creates no added value.

(PCICB, 2002: 1)

It was therefore felt that limiting the respondents to the target group in this particular research down to contract/project manager level was justified, given the potential problems of surveying at the foreman/site agent level or below, as stated above.

PART II

Research instrument

Description of instruments and procedures used

The objective *dependent variables* used to operationalise 'organisational effectiveness' in this research were the success ratings of Hong Kong building contractors employed on public sector housing contracts awarded and operated by the HKHA. These variables were measured (or determined) by means of two interlinked systems operated within the HKHD and used to rate the performance or success of companies on the HKHA's List of Building Contractors, as follows:

- The Performance Assessment Scoring System (PASS) scores over the five-year period from 1997 to 2002 (i.e. preceding, including and immediately following the implementation of the OC survey instrument). This period was chosen as it was only during this interval that the PASS (April 1997 edition) ran under a relatively stable state (i.e. without major systemic changes) and with a relatively steady higher workload level, and therefore the basis for comparison of contractors' ratings was able to be undertaken on a like-for-like basis.
- The 'Premier League System' (PLS) – a scored and ranked matrix developed by the HD to determine its preferred 'quality' partners for future tendering exercises.[2]

The *independent variables* which represented the organisational culture of the respondent companies, categorised as the four traits and 12 survey factors, were measured by applying the Denison Organisational Culture Survey (DOCS) instrument developed by Denison and Neale in 1994. DOCS uses a set of 60 statements which attempt to describe the different aspects of an organisation's culture and derive views on the way in which an organisation operates. To complete the survey, participants indicate how much they agree (or disagree) with each of the 60 statements (i.e. if the statement is a good description of the way that things are typically done in the respondent's organisation) and then he/she should indicate a degree of agreement with that statement. If the statement is not a good description of the way things typically work in the respondent's organisation, then a degree of disagreement should be indicated. The range of ratings available to the respondent is shown in a sample section of DOCS in Table 6.1.

The survey instructions state that 'If you have difficulty deciding between two response options, select the one that you think most closely describes the organization. Neutral should be used when you are stuck in the middle and neither agree nor disagree with the statement' (Coffey, 2002). The consideration of the handling of 'neutral' ratings in surveys carried out among respondents predominantly of Asian ethnicity has been the subject of wide discussion among cultural researchers over the past few years and is discussed in more detail in a later section of this chapter. In order to attempt to maximise the participation by listed companies, the study was first introduced to the industry's representative body, the Hong Kong Construction Association Limited (HKCA). Their full support for the research in general and specifically for the survey to be undertaken among its members (all of the 27 HKHA listed companies forming the research population were also members of HKCA) was obtained and this was considered to be instrumental in facilitating as many construction companies as possible to return completed survey questionnaires (i.e. companies often

Table 6.1 Sample of Denison Organisational Culture Survey (DOCS)

ID	Statement	Level of agreement				
	In this organization .....?	Strongly disagree	Disagree	Neutral	Agree	Strongly agree
1	Most employees are committed to their jobs.	☐ 1	☐ 2	☐ 3	☐ 4	☐ 5
2	When people are provided with adequate information, they can usually make a decision easily.	☐ 1	☐ 2	☐ 3	☐ 4	☐ 5
3	Everyone can get his information as soon as he needs provided that information is shared.	☐ 1	☐ 2	☐ 3	☐ 4	☐ 5
4	Everyone believes that he or she can make a positive impact on the company.	☐ 1	☐ 2	☐ 3	☐ 4	☐ 5
5	Commercial planning is continuously ongoing and everyone will take part in this planning to some extent.	☐ 1	☐ 2	☐ 3	☐ 4	☐ 5

Source: Denison Consulting Group (2009) adapted by Coffey.

tend to be more enthusiastic to support research which has received endorsement from their official representative body). A small pilot study was first carried out to determine how easily the original survey instrument could be used by the target respondents, and copies of the OCS questionnaire were handed to six construction companies via their general managers, who were participating at a quality housing workshop being held during December 2000. The managers discussed the questionnaire with project staff and construction managers within their own company and, based on the feedback received from them, some further revisions to the instrument were made and back-translation procedures reapplied. The finalised version of the OCS questionnaire was eventually promulgated and used.

Administration of instruments and procedures

The research was undertaken based on the following general methodologies:

1 Examination of secondary data (PASS scores, score leagues and PLS performance).
2 Survey based on 'The Denison Organisational Cultural Model'.
3 Statistical analysis to test for correlation in data from both (1) and (2).

4   Undertaking four mini-case studies pertaining to successful and unsuccessful construction companies at the top and bottom of the PASS score leagues (and/or included/not included in the HKHD's Premier Score League) in order to:

- test the suitability of the DOCS as a cultural evaluation tool both for use in Hong Kong and on companies constructing public housing;
- gauge the sensitivity of the DOCS to distinguish the range of performance derived/measured from the secondary data.

Secondary data sources

As mentioned earlier, the PASS – (April 1997) edition of the system was radically amended after a substantive review and eventually reissued as the PASS – (January 2002) edition. This revised version of the system scores many factors based on an absolute 'right' or 'wrong' basis and is somewhat more critically discerning of construction quality performance than its predecessor. As a result, the scores of some contractors who could generally complete construction projects to acceptable standards, but whose management was not strong during the intervening construction pre-completion period were so severely altered during the initial 12-month period following introduction that they were therefore not usable for the thesis research.[3]

The newest version of PASS also post-dates the period of data collection, another reason for not utilising it in this thesis. A summary of the main components of PASS 1997 and its use in determining the relative success outcomes of listed construction companies was given earlier in Chapter 1; however, it is necessary to describe in this section the methodology applied when using the PASS results to make the necessary tests for significant relationships with data obtained from the application of the DOCS. In order to ensure that the secondary data utilisation was consistent in terms of how it was applied to determine the relative success ratings of the companies under study, some basic parameters were established as follows:

1   Because there was a substantial change to both the system scoring methodology and some of the major performance standards with which compliance was being measured by PASS in April 1997, the PASS (April 1997) edition (PASS 1997) has been used as the basic data source. Figure 6.4 illustrates the system conceptually.

Scores drawn from this system have been used to form score leagues to determine tendering opportunities for the five-year period up until March 2002 when PASS 1997 was terminated and scores based on a new PASS (January 2002) edition (a substantially amended system) replaced them from that point in time onwards.

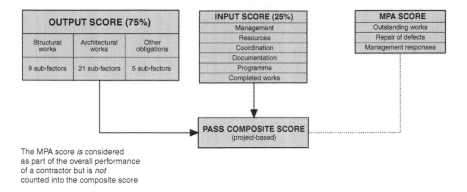

The MPA score *is* considered
as part of the overall performance
of a contractor but is *not*
counted into the composite score

Figure 6.4 The Performance Assessment Scoring System (PASS).

2   All the PASS 'Output' and 'Input' score factors and their related score sub-factors of respondent construction companies have been used, as these holistically represent their performance on all projects being undertaken by them over a two-year duration. The composite score was not used, as this is a measure designed specifically for use in limited three-monthly time periods for making a short-term evaluation of the relative performance of contractors in order to allocate three-monthly tendering opportunities. The maintenance period score was also not used, as this depends on rather varied and subjective reporting by external property management companies on factors such as defects rectification work and completion of outstanding works beyond the construction period. In the past it was used only as an indicator of a contractor's post-construction stage achievements, which is why it does not form part of the overall composite score used for allocating tendering opportunities.

3   The PASS datasets used were derived from the two-year period 2000/2001 as only during this period did *all* companies who responded to the DOCS also have valid performance scores under PASS. The end of this period also immediately preceded the issue out and subsequent return of responses to the DOCS, and so is contemporaneous for the purpose of making the culture – performance relationship analysis.

An example of this score-trend comparison approach is shown in Figure 6.5. The various lines shown in this figure represent the following:

• *Contractor* – The score trend of quarterly PASS scores of a particular contractor in the New Works Group 2 (NW2), i.e. the higher contract value group of the list.

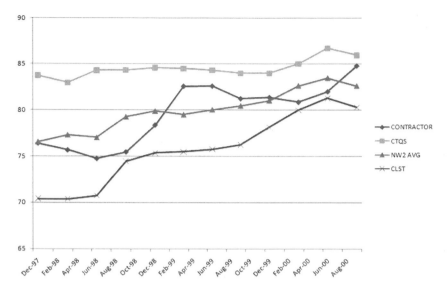

Figure 6.5 Typical PASS score trend comparison.

- *CTQS* – Composite Target Quality Score – the quarterly score of the first quartile of all contractors' scores (i.e. it is the 25 percentile of the contractors' league score).
- *NW2 AVG* – the average quarterly score of *All* NW2 contractors with a score during that period.
- *CLST* – Composite Lower Score Threshold – the quarterly score of the fourth quartile of all contractors' scores (i.e. it is the 75 percentile of the contractors' league score).

The score-trend graphing system was also used as a basis for the performance comparison in the mini-case studies in Chapter 9 where a longer term view of performance was necessary.

While PASS measures objectively the performance of public housing contractors, when considering potential future 'partners' with whom the HKHA can confidently allocate its major new-build construction workload, a conscious decision was taken to consider other factors which represented successful business outcomes, rather than purely basing the consideration on PASS performance scores (however objective) derived largely from measuring compliance/non-compliance with the client's technical specifications. A methodology was developed for the purposes of determining a preferred shortlist of contractors during the latter part of 2000 and the earlier part of 2001 in order to form a ranked list of preferred contractors with whom the HKHA desired to entrust its major future workload. This methodology became known as the Premier League System (PLS) and was based on the following criteria being used to grade performance of all 'active' construction companies undertaking new-works public housing building contracts for the HKHA:

- status (confirmed/probationary);
- number of six-month PASS composite scores below the composite lower score threshold (CLST) quartile;
- total number of 'adverse' reports (ARs);
- total number of performance reports (PRs) (has to be $\geq$ 24 nos);
- Ratio of ARs divided by the total number of PRs (has to be $\leq$ 10 per cent of PR);
- number of instances of 'suspension' from tendering;
- number of current projects (has to be $\geq$ 2 nos).

Measurement of performance in this system was slightly different to that used for generating the PASS score leagues described earlier in this section. Thus, in order to avoid any 'flattening' effect of performance differentiation of companies resulting from simply averaging all of their scores achieved on all projects being undertaken for the five-year period from 1997 until 2002 to one absolute score, the quarterly performance profile of each company over the period was used and compared against a benchmark profile in order to arrive at a final analysis of the overall success outcomes of each individual company. The PLS has, to a lesser extent, also been used as a dependent variable representative of effective or ineffective business performance by the surveyed companies. This PLS league was also considered when analysis of significance of correlation was being undertaken against the DOCS results to ascertain for the benefit of feedback to the HKHD, whether one or the other (i.e. PASS or PLS) correlated more strongly, and the results are reported in Chapter 7.

Primary data sources

The Denison Organisational Culture Survey (DOCS)

The original DOCS (i.e. the actual 'survey' instrument as separate from the original Organisational Culture 'model' developed solely by Denison in 1990) was further co-developed by Denison and Neale (1994) and Denison and Mishra (1995) as a management measurement tool to link organisational culture to tangible bottom-line performance measures such as:

- overall performance;
- profitability;
- market share;
- quality;
- sales growth;

- innovation;
- employee satisfaction.

Based on research drawn from data obtained from over 1,000 high- and low-performing organisations and carried out over a 15-year period, first as the subject of a doctoral thesis at the University of Michigan Business School (1982) and subsequently since that time up until the present day as post-doctoral research and consultancy, Denison found that the following four culture traits can have a significant impact on organisational performance:

- involvement;
- consistency;
- adaptability;
- mission.

Table 6.2 illustrates the differences as perceived by Denison and Neale (1994) between the DOCS instrument and 'traditional' culture surveys.

According to the information available on the 'Denison Consulting' website,

> Traditionally organizational culture surveys have taken a behavioral approach ... making it difficult to link the results back to business. This survey enables leaders, key stakeholders and employees

Table 6.2 Comparison of DOCS and other cultural surveys

Denison model	Other cultural models
• Behaviourally based	• Often psychologically based or personality based
• Designed and created within the business environment	• Often designed and created within the academic environment
• Business language used to explore business-level issues	• Often non-business language which must be converted through interpretation to the business context
• Linked to bottom-line business results	• Often unclear about specific links to business results; little if any research conducted placing cultural elements in relation to performance
• Fast and easy to implement	• Often extensive time required to implement assessment and/or interpret results in the business context
• Applicable to all levels of the organisation	• Specifically designed for either top-level or front-line implementation

Source: Denison Consulting Group (2009).

to understand the impact their culture has on their organization's performance and learn how to redirect their culture to improve organizational effectiveness.

(Denison and Neale, 1994)

The DOCS has 60 items that are designed to measure four 'traits' and 12 'management practices' identified in Denison's research (Denison, 1984) as making up an organisation's culture. When operating the survey, the following methodology is generally employed by Denison's own researchers and/or other organisations in the consulting network and researchers also utilising the tool.[4] Individual surveys (gathered using paper OCR forms or online returns) are collectively tabulated into a graphic profile that snapshots and benchmarks any organisation's culture to that of higher and lower performing organisations in the same, or other, business sectors. Longitudinal use of the survey can then provide a measure of an organisation's progress towards achieving a high-performance culture and optimum performance, when compared to the initial survey. The model has been used to make external comparisons between various organisations for benchmarking purposes, or has been operated internally within organisations to compare inter-group performance and effect change programmes in either. The efficacy of such change programmes, when operated over time within organisations, can then be measured longitudinally in order to gauge the success of the change programmes. More detailed descriptions of previous research carried out using this survey (and other organisational culture survey instruments) may be found in Chapter 4; however, it is worth noting that the instrument continues to be popular among 'culture change' and organisational consultants, owing to its ability to show cultural factors of an organisation in a manner which can be easily understood and assimilated by managers. The fact that the survey process and subsequent analysis and report generation have been largely automated and costs of undertaking such a survey are fairly reasonable when compared to carrying out one-off consultancies means that it is still being widely used in current re-engineering and culture-change business programmes. Denison *et al.* (2000) conducted an investigation of the validity and reliability of the DOCS using responses from 36,542 individuals in 94 organisations. Their paper concludes that the results from the tests undertaken on the DOCS model show 'good support for both the measurement model and the theoretical structure implied by the framework'. An adapted summary of the analysis procedures used in this latter study is drawn from an earlier part-paper by Cho (2000) and has been adapted by the author using information drawn from subsequent research undertaken by Denison *et al.* (2000) and is shown in Table 6.3.

In the earlier research, Cho (2000) also estimated a structural equation model (SEM) for the DOCS indices in order to produce a 'best-fit' model, which provided strong support for the relationship of the model to a variety of organisational performance measures. The 'best-fit' model fulfilled the

Table 6.3 Tests of the validity and reliability of the DOCS model

Refinement procedures	Reason	Methods applied by Cho (2000) or others
Dimensionality	To identify the presence of sub- or super-ordinate factors in the instrument	Factor analysis: both exploratory and confirmatory factor analytic techniques were integral to this process
Item-level statistical analysis	This property is reflected in the empirical finding that scores are highly inter-correlated, i.e. the tendency to endorse one construct indicator should be associated with the tendency to endorse other, alternative construct indicators	Empirical indicators of item consistency include item-total correlations, inter-item correlations, alpha value changes when items are deleted and, in the case of factor analysis, factor loadings
Content-based item-level analysis	To judge whether item-content accurately represents the content domain of the construct	This procedure was applied when the DOCS questionnaire was developed (Denison and Neale, 1994)
Independent sample replication	Thompson (1994) stated: 'If science is the business of discovering replicable effects, because statistical significance tests do not evaluate result replicability." He also stated that actually re-performing the study with a new sample ('external replication') is the only way to directly assess replicability (Kier, 1997)	Item-level refinements were conducted on one sample and then a replicated measure of performance was carried out on a second sample. Replication methods are approximate variance methods that estimate the variance based on the variability of estimates formed from sub-samples of the full sample. The sub-samples are generated to properly reflect the variability due to the sample design
Assessment of discriminant validity	Cases were rated as successful under this criterion if they reported any tests of discriminant validity, including either reports that some factors correlated with a criterion but others did not, or more rigorous, convergent and discriminant validity investigations	

Source: Denison *et al.* (2000), adapted by Coffey.

criteria of the goodness-of-fit measures. The chi-square value of 3177.4 with 44 degrees of freedom at a significance level of $p<0.001$ is regarded as 'extremely significant'. Furthermore, the RMSEA value of 0.044, which is lower than the threshold value of 0.05, also supports the 'good fit' of the model (Hox and Bechger, 1998; Motulsky and Christopoulos, 2003).

Adaptation of the basic survey instrument for the specific research scenario of the thesis

Brislin (1976: 216–217) states that

> the instruments used in much cross-cultural psychological research were developed in one culture (often the middle-class United States). In the development, the researchers take advantage of common experiences shared by people of that culture. Such instruments may have very limited usefulness in another culture in which people do not attach the same value to those experiences.

Brislin (1976: 217) importantly continues,

> a major problem is that researchers use instruments (without modification) in one culture (let's call it A) that were designed, pre-tested, revised, validated, and so forth, in another culture (B). The problem arises when the researcher tries to reach conclusions about culture A by scoring according to the norms derived in culture B. The criticism, of course, is that norms for B may be irrelevant for A, and that the results for such research can be false and misleading.

The described limitation fits exactly the situation faced in the current study by this researcher, namely that the original instrument (i.e. the 'Denison Organisational Culture Survey') was developed in the United States for use among companies in that country and was now being considered for use in Asia, and in particular in Hong Kong. According to Brislin (1976) and several other authors who have, since the 1960s, been studying research issues between cultures (Berry, 1969; Triandis, 1964, 1972; Brislin *et al.* 1973; Price-Williams, 1974), the 'emic/etic' distinction distinguishes the two major goals of cross-cultural research. Briefly summarising Brislin's (1976) definitions, these goals are:

- Emic analysis, i.e. the documentation of valid principles that describe behaviour in any one culture under study, taking into account what the people themselves value as meaningful and important.
- Etic analysis, i.e. the making of generalisations across cultures; that take into account all human behaviour.

- Another concern facing the researcher that arises from the emic/etic debate is that of the 'equivalence' of concepts across cultures.

In this context, the question of *equivalence* refers to whether the concepts being investigated, especially the way in which they are measured, have the same meaning for the culture which devised the instrument as they do for the culture to which the instrument is being administered. Morris *et al.* (1999), drawing on a study of the justice judgement literature, established a synergistic relationship between emic and etic approaches to research by identifying dynamics which allowed each approach to stimulate the other.

This extended previous research and discussion on the merits/demerits of emic and etic insights of culture complemented one another (Berry, 1990; Brett *et al.*, 1997). Morris *et al.* point out that depending in which paradigm they were operating, their assumptions on the nature of culture being studied was different (1999: 782): 'Emic researchers tend to assume that a culture is best understood as an interconnected whole, or system, whereas etic researchers are more likely to isolate particular components of culture and state hypotheses about their distinct antecedents and consequents.' Seen from this perspective, the research described in this book is based primarily on an etic construct; however, in view of the author's personal immersion for over 20 years within the Hong Kong construction scenario as both an operational front-line project manager and as a policy and systems administrator within the HKSARG's public housing sector, it is believed that some 'emic smoothing' of views has as a result been brought to the research stance. In Table 6.4 (adapted by the author from Morris *et al.*, 1999: 783), the main assumptions/goals and associated features/methods of the emic and etic perspectives are clearly related.

Recognising the problems involved in researching across cultures within organisations, Schaffer and Riordan (2003a, 2003b) recommend as a 'best practice solution' the use of what they describe as 'a combined emic-etic approach, or a derived etic approach' (2003a). The authors take the view that

> a derived etic approach requires researchers to first attain emic knowledge (usually through observation, and/or participation) about all of the cultures in the study (Berry, 1990; Cheung, Conger, Hau, Lew, and Lau, 1992) this allows them to put aside their culture biases, and to become familiar with the relevant cultural differences in each setting.
>
> (Schaffer and Riordan, 2003a: 2)

In view of the fact that the DOCS was developed for use in (and has until recently largely been restricted to) North America, the following methodology was used to prepare the survey instrument for use in Hong Kong based on methodologies developed by Brislin (1993), who also

Table 6.4 Assumptions of emic and etic perspectives and associated methods

Features	Emic (inside) view	Etic (outside) view
Defining assumptions and goals	Behaviour as seen from the perspective of cultural insiders, in constructs drawn from their self-understandings	Behaviour described from a vantage external to the culture, in constructs that apply equally well to other cultures
	Describe the whole cultural system as a working whole	Describe the ways in which cultural variables fit into general causal models of a particular behaviour
Typical features of methods associated with this view	Observations recorded in a rich qualitative form that avoids impositions of researcher's constructs	Focus on external, measurable features that can be assessed by parallel procedures at different cultural sites
	Long-standing, wide-ranging observation of one setting or a few settings	Brief narrow observation of more than one setting, often a large number of settings
Examples of typical study types	Ethnographic fieldwork; participant observation along with interviews	Multi-setting survey; cross-sectional comparison of responses to instruments measuring *organisation** perceptions and related variables
	Content analysis of texts providing a window into indigenous thinking about *organisation**	Comparative experiment treating culture as quasi-experimental manipulation to assess whether the impact of particular factors varies across cultures

Source: Morris *et al.* (1999), adapted by Coffey.

Note: *These authors used the word 'justice', which has been replaced in this adapted table with the word 'organisation', which is more appropriate contextually to the subject of the thesis.

describes three potential problems associated with the equivalence of meanings across cultures, which are translational, conceptual and metric. He opines (1993: 78) that 'the starting point for most analyses of complex concepts is that they will *not* be perfect equivalence'. Table 6.5 summarises these problems, defining each type of analysis and stating simply the outcomes expected from them.

Brislin (1993: 79) offers the following detailed solution to the first problem, i.e. that of translational equivalence:

> The back-translation procedure is a good place to start examinations of the equivalence issue. In this procedure, material in an original language

Table 6.5 Summary of three equivalence problems

	Translational	Conceptual	Metric
Description of analysis	Examination of descriptions and measures of concepts as they are translated across other languages	Examination of different aspects of a concept that serve the same purpose in different cultures	Examination of same concept across different cultures assuming that (after proper translation) the same scale can be used to measure the concept
Outcomes	Easily translatable etic aspects can be determined Materials that have to be revised after translation identify emic aspects	First identifies etic aspects and then further identifies emic aspects that are related to the etic in the various cultures under study	If concepts are so similar in different cultures, establishment of metric equivalence is warranted. However, owing to the complexity of human behaviour and the amplification of this complexity when studying behaviour across different cultures, metric equivalence is often not pursued

Source: Brislin (1993, pp 77–89), adapted by Coffey.

is carefully prepared ... a bilingual then translates the material to the target language, and a second bilingual (unfamiliar with the efforts of the first bilingual) translates the material back to English. The two English versions can then be examined to determine what 'comes through' clearly, and the assumption (not yet proven) can be entertained that the target-language version is adequate if the two English versions are equivalent.

Usunier (1998: 51) builds on this, apparently improving the methodology, and states that

A more sophisticated solution to the problem of translation has been suggested by (Campbell and Werner, 1970). Research instruments should be developed by collaborators in the two cultures and items, questions, or other survey materials should be generated jointly. After back-translation or an initial translation process has been performed, there is an opportunity to change the source language wording. This technique, called *decentering*, not only changes the target language, as in the previous techniques, but also allows the word and sentences in the source language to be changed if this provides enhanced accuracy.

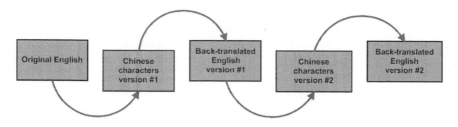

Figure 6.6 Back-translation methodology.

Brislin (1976: 223–224) defines decentring as 'the research project is not centred around any one culture or language. Instead, the idiosyncrasies of each language under study contribute to the final version of the questionnaire.' Triandis (1972, 1976) outlined procedures involving the development of suitable research instruments within each culture, i.e. there is no translation of material from one language to another. Brislin (1993) suggests that the disadvantage of the latter is that the etic may be obscured owing to the development of purely emic instruments in each culture. Usunier (1998) concludes that 'When two languages and cultures present wide variations, such as Korean and English, combining parallel and back-translation techniques provides a higher level of equivalence.'

Based on this previous research, it was decided to utilise the back-translation methodology in this thesis for the following reasons:

- The original instrument is extremely robust and, due to being used in a large number of surveys over a long period of time, did not require significant alteration, as to do so would have negated the basis for comparisons to be drawn between this research and all that has been done before.
- Figure 6.6 illustrates the flow of the specific back-translation path applied to this study. Owing possibly to Hong Kong's continued attachment to British-style tertiary educational courses related to the building professions and the continued use of English contract documentation in the construction industry, English is still widely spoken and used by middle- and upper- level managers in construction companies, and the Chinese character version #2 was used merely to back up and supplement the back-translated English language version #2 of the instrument, in the event that a 'prompt' on exact meaning might be required. There was thus no need to engage in the more rigorous, 'parallel' or 'mixed translation' methodologies.

Having thus prepared the survey instrument, Version 1 (known as the back-translated English #2) was forwarded to four senior managers in major construction companies for their comments on the general suitability of the survey for their sector of the industry and in terms of the cultural/ cross-cultural issues. Based on the feedback obtained from this exercise, some minor modifications were made to certain questions in the survey as a result of this feedback. An interesting phenomenon was noted when the first back-translated version was received in that the suggested modifications by the translator related to a perceived need to 'de-Americanise' certain phrases and idioms in some of the questions. An example of this was 'In this organisation, problems tend to fall through the cracks' which was eventually translated into 'In our company, many things which are important go unnoticed'. For another questionnaire statement 'Work is organised so that each person can see the relationship between his or her job and the goals of the organisation', the revised wording after decentering and back-translation became 'Jobs are allocated so that people can realise the relationship between their job and those of others and have a better understanding of the company's goals'. Survey Version 4 (English and Chinese) was eventually the iteration of the instrument used to obtain the data from respondents. A copy of the final version of the questionnaire is shown in Appendix B at the end of the book.

Consideration of potential cultural bias in responses to the survey instrument

Several studies have examined whether responses to questionnaires are significantly affected by the inclusion of response scale mid-points. Spagna (1984) found that by omitting the mid-point, the frequencies of the adjacent scale responses were generally increased. Si and Cullen (1998) conducted a study using management research instruments developed in Western countries to ascertain whether the use of the response scales within such instruments would significantly affect results derived from these questionnaires when they were administered to respondents in other cultures. Their basic research hypothesis was that results obtained from questionnaires using Likert-type scales, when administered to Chinese-Japanese-Hong Kong (CJH) managers, would differ from those obtained from their Western counterparts responding to the same surveys. Specifically, these authors anticipated that response variations would be minimised by CJH respondents choosing mid-point responses where questions had odd-numbered response categories. The basis for their hypothesis originated from a view that Confucian teachings regarding the correctness of always trying to follow a 'middle way' administered to CJH managers during their education and in their social and family lives represents an ingrained philosophy ('an unconscious and automatic neutral response') which might significantly affect their responses to questionnaires with variable response categories that require decisive or value judgements. This view supports earlier findings by other culture-based researchers (Hofstede and Bond, 1988; Hofstede, 1991) that also concluded that Asian managers following a Confucianist logic/ethic were biased to select the 'middle-way' response, not only when faced with responding to questionnaires but choices in life generally. Important to note in respect of this thesis is that although the findings of Si and Cullen's (1998) research *did* appear to generally support the effects of this 'central tendency' trait, the authors point out (1998: 226) that 'managers from Hong Kong and Japan have more exposure to Western management practices than their PRC counterparts and this may point to differences in tendency to avoid extremes of response not being so marked as a result'. The authors also discovered that these 'middle-way' responses did not affect the overall mean response levels. Other researchers have had opposing results. Dawes (1972), for instance, found that the use of a scale mid-point resulted in fewer positive responses to specific questions, whereas Garland (1991) concluded the opposite, i.e. a mid-point resulted in fewer negative ratings. The literature then indicates that there are mixed views on this issue and the choice falls to the researcher to decide on the risk involved in using mid-point scales in varying response category questions.[5]

In order therefore to be able to offer the results obtained in this research back to the central data pool held by the Denison Consulting Group, a five-point Likert response scale in the adapted DOCS has been retained and, as may be seen in the later research findings, this did not indeed result in any problems concerning the eventual responses to the questionnaire.

Use of Denison Organisational Cultural Survey

There were 53 building contractors in total on the HA's list in the Building (New Works) Category. The companies are subdivided by group, based on the maximum permissible contract values able to be held, as follows

- *Group NW1* (26 nos) – Contractors are eligible to tender for contracts with a value of up to HK$450 million.
- Group NW2 (27 nos) – Contractors are eligible to tender for contracts of unlimited value.

The modified DOCS forms were issued to *all* 53 companies in the two groups referred to above and in the appropriate category on the list. Some of the companies on the list were at the time of survey 'inactive' or 'suspended from tendering', and therefore it was known that it would be unlikely that such companies would respond. It should be noted however that not all of the survey returns actually received were eventually used due to this not meeting the basic criteria laid out in the instructions for filling in the survey questionnaire.

Methodology of application of the qualitative methodology

Qualitative research techniques described later in this section were applied to three of the Premier League Scheme (PLS) contractors as well as to one non-PLS company, the latter having both exceptionally poor PASS scores and overall performance. The qualitative research techniques were used to provide further depth to the examination of the relationship between organisational culture strengths/weaknesses and construction sector business success outcomes. Qualitatively founded research was conducted on two of the top, one of the middle and one of the bottom companies, based on this PLS matrix, in order to establish whether there were any other underlying reasons for their performance (measured by the PASS) being different from one another and also whether this performance difference was entirely based on measurable differences in their organisational cultures. It was also considered to be fundamental to examine in more depth the cultural traits of the best of the most successful companies against one of those considered to be least successful and to determine whether any other factors were operating within the companies which would explain the differences in performance between both groups. The research within these four companies consisted of the following:

- mini case studies;
- examination and analysis of open-ended and performance self-rating questions and sub-questions added to the DOCS instrument.

The results of this research have been written up as four mini-case studies (covered in Chapter 9).

Limitations of the methodology

HA List of Building Contractors as the survey population and use of PASS-generated performance results

There were some limitations to the methodology and these were that the DOCS was issued only to those companies that fulfilled the following criteria:

- Only the 53 companies on the Hong Kong Housing Authority (HKHA)'s List of Building Contractors, Building (New Works) Category, Groups NW1 and NW2 have been used as the survey population. This is because only this group of companies has its performance objectively measured by PASS. Other contractor lists (e.g. maintenance, shopping centre improvements) do not operate to standardised building designs or have objective systems operating to measure performance of contractors.
- They had to be currently carrying out (or have carried out) work during the period April 1999 to March 2001 for the HKHA on at least one new works building project in the NW1 or NW2 group value ranges (in other words, they had active or historic PASS profiles; otherwise no comparisons would be possible between DOCS and PASS results).

- Only performance on HKHA projects was considered due to the fact that other government contracts are operated on a wide variety of building types with dissimilar performance monitoring systems.
- PASS data used were based solely on the 'output' (quality of constructed work), 'progress' (against contract programme) aspects of performance and on the 'input' (management) aspects. The former are objectively and accurately scored against the provisions of the contract documents, including (but not limited to) specifications, drawings, architects' instructions, etc., while the latter, although somewhat more subjectively scored, are based on a mixture of professional judgement and a series of guidelines to ensure consistency of assessments.
- The numerical scores used to determine comparative success ratings of companies have been drawn from the Hong Kong Housing Department (HKHD)'s PASS 1997 edition and covers performance for the period April 1997 to March 2000. This is because PASS remained unrevised and stable during this period, having undergone major revisions in 1997 and again in 2002.

Limitations of response during survey administration

The original survey invitation sent out in May 2000 requested respondent companies to provide six to nine returns based on representing the following target managers' groups:

- Chairman/CEO/Managing Director – *one response*;
- General Manager (Construction/Building) – *one response*;
- Contract Managers (Construction/Building) – *two to three responses*;
- Other managers associated with construction operations (e.g. Quality/Safety/Estimating/Materials Purchasing, etc.) – *two to four responses*.

The survey invitation was accompanied by a soft copy of the English and Chinese versions of the questionnaire and also advised potential respondents that they could fill in the survey electronically via a secure, temporary website specifically set up for the purpose. This website was subsequently closed one month after the final promulgated cut-off date for the return of surveys. Over several weeks (eventually months) following the dispatch of the first survey requests, reminders were sent out to companies either to request non-returnees to provide returns, or to ask for further returns from companies who had responded, but whose total number of completed survey forms was only one or two, considered unusable or unrepresentative as a bona fide response for data analysis purpose. This process proved extremely time-consuming and eventually the survey was closed some six months after it was commenced. Having spoken to various respondents over the course of the survey being lodged with potential respondent companies, the author gathered that the procrastination of some companies was due mainly to the following two reasons:

- The numbers of individual surveys (up to 10 or 12 a week) that construction companies were sent by the various tertiary educational institutes both in Hong Kong and elsewhere.
- Companies who were not actively working for the HA or who had poor performance scores were reluctant to join in the research study as it was felt that their indifferent performance/non-activity might in some way be highlighted, presenting them in a negative light.

The respondents questioned who did eventually submit late responses indicated that it was only after several letters, phone calls, emails and faxes had been received that they took the time to read the contents of the invitation more closely and decided to participate, particularly as they saw the eventual outcome (i.e. a benchmarking report on their company's cultural comparison) being of some commercial benefit to them.

As previously mentioned, the total number of NW1/NW2 listed construction companies to which surveys were dispatched was 53 and the total number of returned survey sets was 29 (54.7 per cent) and, based on the fact that four returns appeared to represent a reasonable cross-section of a company's site, middle and higher management, of these some 23 sets were usable in the study data set (43.40 per cent). However, one of these firms did not have an active project during the two-year performance period and was discarded, meaning that the eventual number of companies submitting usable returns was 22 (41.50 per cent). The detailed results of return of surveys, together with basic demographics of respondents, are shown in Chapter 7.

PART III

Data analysis methodology/administration

Basic survey data and demographics

Incoming surveys basically fell into three categories and each of these categories was handled in the following manner:

1 *Paper format* – These were coded with a confidential identity (ID) number extracted from a secure computer file relating IDs to company names held in a secure location on the researcher's computer. Each company was given 12 separate ID numbers to ensure sufficiency of cover for the maximum number of anticipated returns, i.e. up to 12 questionnaires. The survey results were then input manually by a

confidential typist into a secure *Microsoft (MS) Excel* data spread sheet, specifically set up for the purpose in an internet-connected computer (however, this computer was not connected to any Local Area Network (LAN) system and so access to survey information even by others within the researcher's own organisation was not possible).

2 *Diskette and email attachment format* – These were also coded with ID numbers and survey results were then electronically transferred into the same *Excel* data spread sheet.

3 *Electronic web-based format* – returns via the internet site were automatically coded and transferred into the same *Excel* database. Automatic email receipts were also generated to respondents thanking them for their participation.

The raw data for the 60 basic questions in the questionnaire which were to form the basis of the company OC profiles were transferred electronically from the *Excel* software data spread sheet into an *SPSS version 11.0* (SPSS, Inc., Chicago, ILL.) data editor file for further analysis. Qualitative data were not entered into this database and were analysed separately as detailed later in this chapter. Age, salary range, ethnicity, company position and experience and basic descriptive statistics (e.g. means of responses to individual survey questions, standard deviations, medians and quartiles) were analysed using *MS Excel* software.

Treatment of PASS data

The PASS data used at the time of the DOCS being undertaken rested within the 'live' section of the HKHA's main frame computer bank.[6] The relevant data were part-manually input (1999 data) by a confidential typist and part-electronically (2000) transferred into the *SPSS version 11.0* (SPSS, Inc., Chicago, ILL.) data editor file for further analysis.

Basic data analysis methodologies used

Statistical analysis – general description

Basic factor analysis was carried out to test the strength of the application of the DOCS in the specific research setting in terms of measuring the underlying dimensions of organisational culture. Cronbach's alpha analysis was used to determine the coefficient of reliability of the dataset, and correlation and multiple linear regression techniques were used to examine and compare the quantitative datasets obtained from DOCS and PASS. The significance level applied to all data analysis was set at 95 per cent for two-tailed normal distribution. The DOCS questionnaire was based on a Likert-type five-point scale to provide a continuous distribution in the survey.

Reliability and validity of DOCS

It is desirable that any measurement instrument possesses reliability and validity and, for multi-item measurement instruments such as DOCS, any items which do not contribute to the instrument's reliability and validity should preferably be removed. Mention was made previously regarding the validity and reliability testing of the DOCS by Cho (2000) and Denison *et al.* (2000). Clearly, despite the fact that much work has already been carried out by these researchers, the fact that the DOCS has been specifically adapted in this thesis for use in Hong Kong and has been operated in a particular industry and sector warrants further comment on the validity and reliability of the adapted questionnaire based on the response data collected for the research study being described in this book. Findings from the results obtained that impact on the consideration of various types of validity are detailed in Chapter 7. However, some discussion on the testing undertaken on the DOCS by other researchers is given in the following sections of this chapter.

Reliability

The degree to which items in a data measurement set are homogeneous is referred to as the internal consistency of the set and can be estimated using a reliability coefficient such as Cronbach's alpha (Cronbach, 1951). Among any set or subset of items, computation of the reliability coefficient using Cronbach's methodology, it is possible to identify the subset having the greatest degree of reliability. According to Nunally (1970), typically coefficients of 0.7 or more are considered adequate to demonstrate factor reliability, and this is supported by Cox *et al.* (1998) who go on to suggest that, for a relatively small number of items, an (α) value of greater than 0.61 is generally acceptable. An internal consistency analysis was performed using this methodology separately for each of the 12 critical factors of organisational culture derived from the Denison model (i.e. the index items) and also the four overriding critical factors (i.e. the cultural traits). The results are shown in Chapters 7 and 8.

Validity

As mentioned earlier in this chapter, validity of a measure refers to its ability to measure what it was supposed to measure. It is necessary to consider three types of validity; in fact, an ideal validation includes several types of evidence which span all three of these categories, which are as follows:

- content validity;
- criterion-related validity;
- construct validity.

Before examining these methodological issues specifically in relation to the research study under consideration, it is worthwhile to briefly examine what various writers and researchers have had to say about the concept of validity, and to note that care is required in this area of establishing methodologies. Trochim (2001) tells us:

> There's an awful lot of confusion in the methodological literature that stems from the wide variety of labels that are used to describe the validity of measures. I want to make two cases here. First, it's dumb to limit our scope only to the validity of measures. We really want to talk about the validity of any operationalization. That is, any time you translate a concept or construct into a functioning and operating reality (*the operationalization*), you need to be concerned about how well you did the translation.

Schmidt and Hunter (1977) conducted some landmark research, which provided a different explanation for the differences that were being observed in validity coefficients. They showed that the variability in validity results appeared not to be as was originally thought (i.e. due to real differences between settings), but was caused by artificial error sources. They identified three such error sources that accounted for the majority of variation in validities across studies. These were:

- sampling error;
- differences among studies in the amount of test range restriction;
- differences among studies in the reliability of criterion measures.

Importantly, this meant that in many cases the 'situational specificity hypothesis can be shown to be false' (Schmidt and Hunter, 1977). These authors also devised a set of quantitative procedures called meta-analysis which they used to demonstrate that true validities are consistent from situation to situation and that validity is 'generalisable'. DSS Research, a Texas-based market research consultancy (DSS, 2004), note that even if the sampling frame for respondents has been carefully handled and successful Cronbach's alpha tests applied, together with confirmatory or discriminatory factor analysis, the researcher may still not produce a measure which has construct validity. These authors assert that consistency is necessary, but is not sufficient for construct validity. The researcher must also determine the extent to which the measure correlates with other measures designed to measure the same thing and whether the measure behaves as expected. Evidence of the convergent validity of a test or measure is provided by significant correlations with other approaches to measuring the same construct. With these comments in mind, the research for this thesis handled the validity issue as described below.

Content validity

The content validity of any measure is related to what degree a particular researcher creates a metric that covers the content domain of the variable being measured and that is usually (subjectively) judged by the researchers themselves (Badri *et al.*, 1995). Denison and his various collaborators based their original model of organisational culture and its subsequent development on a series of robust literature reviews and they have replicated their results (i.e. the strength of the relationship between organisational culture and organisational effectiveness) on many occasions, and these studies have been closely examined in Chapter 5 (this volume). Since its widespread publication in 1994, other researchers, either separately or in collaboration with Denison, have also used DOCS to demonstrate the relationship between organisational culture and organisational effectiveness (Denison *et al.*, 1996; Cho, 2000; Fisher, 2000; Fisher and Alford, 2000; Fey and Denison, 2003), and this would appear to fully satisfy content validity requirements.

Criterion-related validity

This is sometimes described as predictive or external validity and is most connected with the degree to which a particular instrument is related to, and an independent measure of, the relevant criterion (Badri *et al.*, 1995). Thus the 12 organisational culture items and four cultural trait measures applied to different Hong Kong construction companies will be deemed to possess 'criterion-related validity' if both *collectively* and *individually*, they are significantly positively correlated with effectiveness measures derived from the HKHA's PASS scores and PLS rankings. Denison's various studies mentioned earlier indicate conclusively that DOCS does possess strong validity in this area. Results achieved in the research study reported in this book are discussed in detail in Chapters 7 and 8.

Construct validity

As described earlier, a network of hypotheses was developed to measure the strength of the relationship between organisational culture and effectiveness, and these rely on a different set of non-financial measures of effectiveness to those used previously in the studies by Denison and others. According to Zickmund (2003: 344), 'In its simplest form, if the measure behaves the way it is supposed to, in a pattern of intercorrelation with a variety of other variables, there is evidence for construct validity.' The construct validity of each critical factor measure (i.e. the 60 items of the DOCS), when compared with both the financial and non-financial variables applied in previous research and in this thesis, supports the construct validity of the DOCS instrument.

Identifying underlying factors

The term *factor analysis* was first introduced in the early 1930s by Thurlstone and is a technique used to simplify the numbers of tests and measures with which the researcher is required to cope when examining large quantities of data. In essence, factor analysis is a technique for finding a small number of underlying dimensions from among a large number of variables. The main applications of the technique are to *reduce* the number of variables and to *detect structure* in the relationships between variables. In relation to the DOCS, on the basis that the instrument designers have built up the DOCS from sound previous research and the variables have been validated and found reliable on a huge number of organisations and individual respondents over the 15 or more years of their use, plus the fact that each set of five variables informs one of the 12 critical factors of the DOCS model, a decision was made to retain the results of all 60 variables for the subsequent research investigation. Factor analysis was however carried out using data derived from the current research in order to revalidate the DOCS and to test further for any changes to the underlying dimensions of the instrument. Since the Denison Organisational Culture Model groups the 60 questions (or variables) into four cultural traits, four factors were specified in the factor analysis. A suitable factor loading was selected as the cut-off value and, once a set of common factors had been identified, various factor rotations were tried in order to ascertain the minimum number of significant underlying dimensions that were encapsulated by the operation of the DOCS in the particular research population being examined for this thesis. The results of this factor analysis are reported in detail in Chapter 7, together with opportunities for further research using more advanced factor analysis techniques and implications for the further development of the Denison model.

Summary of data treatment and data analysis methodologies used

The data acquired from the DOCS and PASS, both quantitative and qualitative, and the procedures and tests used to analyse them are shown in Table 6.6.

Chapter summary

This chapter is large in comparison to others in the book but deservedly so, since it contains information that really needs to be read as a whole, and hopefully splitting it into three parts helps the reader makes sense of all the information captured within it. The chapter has attempted to introduce and justify the underlying philosophical paradigms and the major procedures and data-collection methodologies followed before and during the research

Table 6.6 Treatment/analysis of DOCS data

Treatment of data	Methodology
Line item scores calculation	For each of the 60 questions, the corresponding line-item score is assigned as: Strongly disagree: 1, Disagree: 2, Neutral: 3, Agree: 4, Strongly Agree: 5 Reverse scoring is made for the reversed coded questions. A line-item score of a particular contractor is calculated as the mean of scores by all respondents from that contractor.
Index scores calculation	A cultural index is made up of 5 questions. For example, empowerment is made up of the first 5 in the 60 questions. Each of the 12 index scores is calculated as the mean of the constituent line item scores.
Cultural scores calculation	Each of the 4 cultural traits (involvement, consistency, adaptability, and mission) is made up of 3 cultural indexes. They are calculated as the mean of the constituent index scores.
PASS scores calculation	Quarterly PASS scores in structural plus architectural works, other obligations, site management and progress in 2000 and 2001 of the surveyed contractors are employed in this study. For each of the above 4 aspects of PASS, the mean of quarterly PASS scores in the two years is calculated to represent PASS performance of the contractors.
Internal consistency of measures of the DOCS	Based on the whole set of returned questionnaires, internal consistency of measures of the DOCS is assessed by calculating Cronbach's alpha for each index composed of 5 line items, each cultural trait composed of 3 indexes, and all the 4 cultural traits in combination.
Correlation between cultural scores and PASS scores	Based on the cultural scores and PASS scores of the 23 responding contractors, correlation is investigated between cultural and PASS scores. Pearson correlation and linear regression analyses are carried out between each of the cultural scores and each of the PASS scores.
Open-ended questions concerning self-perceptions of 'performance', etc.	Responses to the open-ended questions were tabulated into a simple spread sheet and scores were allocated to each response set and a basic descriptive analysis was applied to them to give some comparison between companies and also to allow more detailed comparisons in the mini-case studies (Chapter 5).
Open-ended questions concerning self-perceptions of 'value'.	Responses to the two-open ended questions were content analysed and categorised and a simple concordance analysis (e.g. 'indexing', 'counting of word frequencies', 'comparison of different usages of a word', 'analysis of keywords' and 'locating of phrases and idioms') was undertaken.
Determination of linear combination of variables in DOCS	All data derived from the 60 main DOCS questions were factor-analysed and results rotated to determine the minimum number of significant underlying dimensions that were operated by the instrument.

Source: Coffey (2003).

described here and later in this book, in order to provide assurance of their appropriateness and reliability. The limitations of the research were explained and the focused parameters of the research topic area were clearly established. The next chapter follows on and describes in detail the outcomes of the data-collection process, and summarises some preliminary findings and subsequent detailed analysis of those data and the results obtained.

Notes

1 This meant that a rich data resource was available for use in the research from the interviews carried out with these three companies before their inclusion into the six-contractor group forming this prestigious league of top-performing construction companies.

2 There were also some subjective self-rated performance data collected from the survey respondents, similar to those used by Denison in the self-assessment open-ended questions within the DOCS to ascertain views of companies on the 'effectiveness' of their organisations. These data were compared with PASS results and are therefore not included as a 'variable' in the sense of other performance/success measures.

3 This is not a reflection on inaccuracy or inadequacy of the 1997 version of PASS; it merely reflected some change in the client's (HKHD) tightened criteria for acceptable construction quality. Contractors whose scores were most subject to change during the transition were not those at the extremes (i.e. 'best'/'worst') but rather those grouped around the middle of the PASS score league. This was in fact one of the major reasons for altering PASS, i.e. to better differentiate the more mediocre performers.

4 Since 1996 Denison and Neale have operated the DOCS under a consultancy, i.e. Denison Consulting.

5 The original DOCS instrument utilises a five-point Likert response scale and in an exchange of emails between the author and Dr Denison in 2000, a view was proffered by Dr Denison that all users of the DOCS, whether corporate or academic, could only have their results subsequently added into the central DOCS database of Denison Consulting if they kept to this five-point response scale. Dr Denison also suggested that in his experience, the potential central tendency rating problem among managers with delegated authority within organisations had not been an issue previously over the years the survey has run.

6 These data were archived in January 2002 once the new version of PASS was promulgated, although official paper archive copies of the data are held in the unit filing system managed by this author for the purposes of referral and are retained for up to 10 years after projects are complete in compliance with limitation clauses of the building contracts operating on such projects.

7 The Hong Kong experiment
Presentation of demographic data, overall results and some descriptive and qualitative analysis

Chapter introduction

Chapter 6 described in its three parts the basic research questions investigated in the context of this book, the establishment of the research instrument used primarily to collect organisational culture data from Hong Kong construction companies and the methodology of its administration and the subsequent analysis methodology. This chapter now presents the data and results obtained from examining the secondary data from the Performance Assessment Scoring System (PASS) scores, held by the Hong Kong Housing Department (HKHD) and from the application of the revised Denison Organisational Culture Survey (DOCS), together with the findings from these data sources. The chapter is generally organised around two major topics: the pattern of results and the analysis of the results relevant to the research questions described in Chapter 6.

Subjects

General data received in the surveys

The previous chapter mentioned that the total number of NW1/NW2 (26/27) listed construction companies to which surveys were dispatched was 53. The total number of returned survey sets was 29 (54.7 per cent) and of these some 23 sets were eventually usable in the study dataset (43.4 per cent), based on post-return established cut-off criteria. The detailed results are shown below.

- *New Works Category Group NW1 Companies* – Ten survey sets were returned from the 26 companies targeted. Of these, eight sets had returns at, or above, four questionnaires and so in terms of usable company returns this represents a return rate of 30.77 per cent. The explanation for the relatively low number of returns appears to be that, in the view of the author, only 11 of the NW1 companies actually had 'active' building contracts with the HA at the time of conducting the survey and therefore enthusiasm to join in the exercise was not high among these companies. Some of the reasons given when the author contacted these companies were:

> 'It is not worthwhile for us to join in, as no comparison can be made between our organisational culture profile and our success rating measured by PASS.'

> 'Our organisational culture will look very weak compared against the larger NW2 firms.'

- *New Works Category Group NW2 Companies* – Nineteen out of the 27 companies targeted returned survey sets. Of these, 15 sets had returns of four or above questionnaires and so in terms of usable company returns this represents a rate of 55 per cent. Apparently, the main reason that these companies gave a much higher return rate than their NW1 counterparts, in the view of the author, was that 20 of the NW2 companies providing usable returns had 'active' building contracts with the HA at the time of conducting the survey and therefore the enthusiasm to join in the exercise was much higher among these companies. The one 'active' company that did not respond was in a situation of undergoing suspension from tendering for all new works building projects for the Hong Kong Housing Authority (HKHA), and had more or less scaled down its organisation in view of reduced workload.[1]

  It is also fair to say that the NW2 companies have better developed management systems, a larger number of managerial and administrative staff to handle the completion of questionnaires and surveys, and also a greater vested interest in taking part in research studies, particularly when they would be able to benefit from the results of the information eventually emanating from this study.

Detailed results

General statistics (figures drawn from all respondents, i.e. not reduced to number of usable responses only)

- *Age distribution* – There were 155 respondents (two had no response) and they were distributed as shown in Figure 7.1. Nearly half of the respondents (49.68 per cent) were in the 40 to 49 age range and 91.62 per cent of respondents were between 30 and 59 years old.
- *Gender* – There were 155 respondents (three had no response and five preferred not to respond) of which 95.97 per cent were males.
- *Educational background* – There were 157 respondents of which 56.05 per cent had degree-level qualifications and 31.21 per cent held

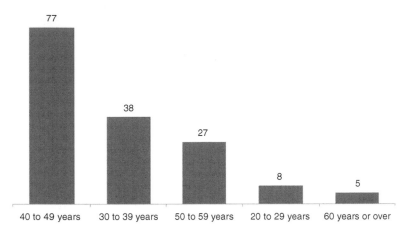

Figure 7.1 DOCS respondents – age.

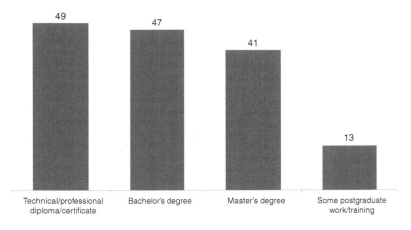

Figure 7.2 DOCS respondents – gender.

some kind of technical/professional diploma/certificate. The respondents were distributed as shown in Figure 7.2.

- *Location of department of work* – There were 155 respondents (two had no response) of which 65.16 per cent were located in construction or general management departments. The respondents were distributed as shown in Figure 7.3.
- *Organisational level* – There were 155 respondents (two had no response) and 65.38 per cent were at middle or senior management level. This particular group within companies was deliberately targeted for response, as explained in Chapter 6, due mainly to the difficulties of obtaining meaningful responses from the multi-layered subcontractors

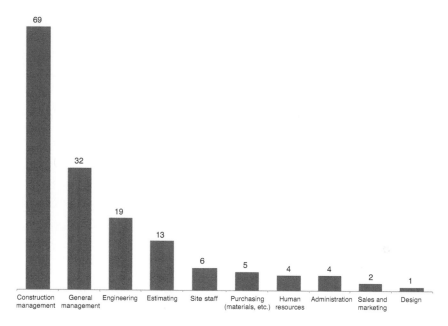

Figure 7.3 DOCS respondents – work department.

and their foremen and gangers. The respondents were distributed as shown in Figure 7.4.

- *Years with organisation* – There were 155 respondents (two had no response) and 65.61 per cent of respondents had worked with the company for six to more than 15 years, while only 6.37 per cent had less than one year of employment with the company. The respondents were distributed as shown in Figure 7.5.
- *Basic salary (annual) in HK dollars* – There were 148 respondents (nine had no response) and 75.66 per cent earned between HK$350,001 to 1,500,000 while only 9.66 per cent earned either more than HK$1,500,001 or less than HK$250,000. The respondents were distributed as shown in Figure 7.6.
- *Ethnic background* – There were 155 respondents (two had no response) and 93.55 per cent were Asian and the remainder were predominantly Caucasian with one Hispanic respondent.
- *General respondent population* – Based on the above data, the predominant population of the survey may generally be described as follows:

  – between 30 to 49 years old;
  – male;

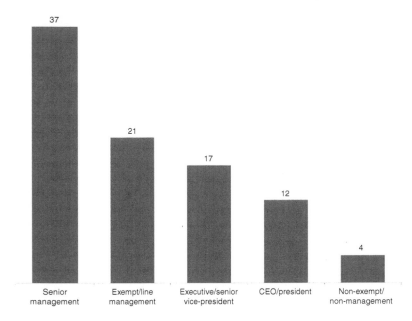

Figure 7.4 DOCS respondents – organisational level.

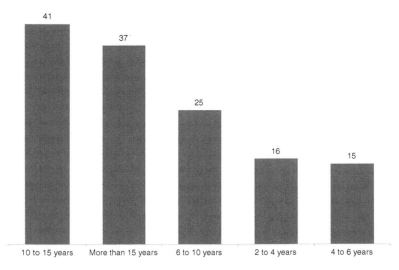

Figure 7.5 DOCS respondents – years with organisation.

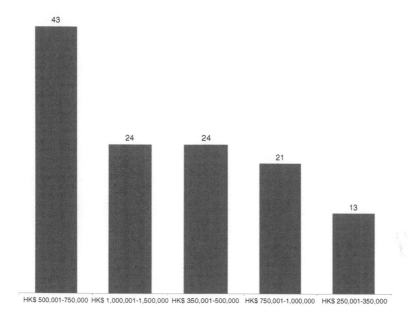

Figure 7.6 DOCS respondents – basic salary (HK$).

- holding a technical/professional diploma/certificate, bachelor's or master's degree;
- senior level (or middle level), general (or construction) managers;
- from six, to more than 15 years with the organisation;
- basic salary between HK$350,001 to 1,500,000;
- with few exceptions, of Asian ethnicity.

Observations regarding data return rates

It is pertinent at this point to give some commentary on the return rates for the survey, as there are some very specific reasons for lower returns in the two contractor groups (i.e. NW1/NW2). Based on a long-standing (i.e. 22 years) experience of working with the Hong Kong construction industry in general, a daily working relationship with companies on the HA list and an in-depth knowledge of all of the respondent contractors, I would offer the following observations on why different companies did/did not choose to respond and also why some companies provided a larger and some a smaller number of responses:

- Those contractors in Group NW1 of the list, because of their inclusion in the group of 26 contractors only able to undertake projects with a contract value of under HK$450M (of which there are only one or two at the most each year), generically tended not to have any 'active' HA projects. These companies may have been reluctant to return surveys, since as they could not see any benefits to themselves from the subsequent data analysis/correlation exercise because they would have no available PASS scores on which to base a comparison. Even after a direct approach was made to such companies to indicate that completion of the requisite number of survey forms would give them a company culture profile which could then be gauged against other similarly obtained profiles, forming a useful comparative tool for strategic consideration of their organisation in relation to future work for the HA, there was still a reluctance to undertake the survey.

- Some of the eight 'probationary' contractors in Group NW2 who were only eligible to undertake *two* Group NW2 projects possibly exhibited an attitude similar to those in Group NW1.
- All HA listed contractors are made aware of their PASS scores and score league positions on a quarterly basis and it was felt that those Group NW2 contractors who did have projects with PASS scores regularly below the CLST (i.e. those falling into the lower score quartile) were reluctant to draw attention to any deficiencies revealed by the results from the OC Survey (an assumption which has been backed up by some subsequent discussion with companies falling into this group). Once again, such contractors were directly approached and the potential benefits of joining in the survey were carefully explained to them, i.e. the ability to be able to benchmark strong and weak cultural traits among the most successful firms in the sector. This approach did in fact persuade a few firms to complete survey returns or increase a previously lower rate of return. This apparent change of heart/rekindling of enthusiasm may have been due to the more 'strategically aware' or commercially experienced managers realising that they were being offered a potentially rare opportunity to benchmark themselves against others in their field, using a new tool which was being administered by an independent researcher who was not materially, competitively or commercially involved with their industry.
- Some contractors explained that they received so many requests to complete surveys or provide information from external sources (e.g. academics/students from tertiary educational institutes, consumer organisations) that they could not devote sufficient resources to the task of completing several (or in some cases any) survey forms. Once again, when directly approached, a few firms did make an extra effort to bring their total number of returns up to four in order to be a part of the whole research exercise and gain feedback from their participation.

Some of the reasons given when the author contacted these companies for non-response were:

> 'We have wasted a lot of time filling surveys before now and have been really disappointed that absolutely nothing of benefit to our company ever comes out of them.'

> 'Our company is proud of its high technical competency and we believe that it is this that is responsible for high PASS scores.'

> 'We don't have the time for our managers to fill in this questionnaire.'

> 'We think our head office culture is pretty good but we can't control the culture of our domestic subcontractors and they are the front-line guys really being marked by the PASS scores.'

These comments are to a degree revisited in Chapter 10 when the research conclusions are discussed, but as far as this chapter is concerned it should be noted that despite such observations and expressed views, 10 Group NW1 and 19 Group NW2 (of whom two were 'probationary') *did* submit survey returns and the above observations are therefore generally directed at those who did not submit a return at all.

Statistical analysis

Clarification of 'levels of significance' used in the data analysis

In order to clarify for the reader the term 'significance' used in the data analysis in this thesis, it is defined in context below.

The *Significance* of results can be reported in various ways; for example, Motulsky (2004: 11), the author of a statistical analysis software package, in an updated webpage definition uses four terms as shown in Table 7.1. These terms, according to the author, are 'not entirely standard' and this book uses the more conventionally accepted definitions provided by Coolican (1990: 174) that are also very similar to the definitions of significance generated by the *SPSS version 11.0* (SPSS, Inc., Chicago, ILL.) software which,

Table 7.1 Definitions of statistical significance

P value	Wording used	Summary symbols
> 0.05	Not significant	ns
0.01 to 0.05	Significant	*
0.001 to 0.1	Very significant	**
< 0.001	Extremely significant	***

Source: Motulsky (2004, p. 12).

based on the relevant probability level, describes significance levels in the following three ways:

- significant: $0.05 > p < 0.01$;
- highly significant: $0.01 > p < 0.001$;
- very highly significant: $0.001 > p$.

All probabilities have been reported based on two-tailed tests, as each comparison had two possible directions.

Descriptive statistics

The descriptive statistics for the 12 organisational culture indices derived from DOCS and used in the correlation and linear regression tests are shown in Table 7.2. From this table, it may be observed that of the 12 cultural indices, the means for 'team orientation' and 'core values' were the highest, indicating that the 157 respondents rated these factors as being important to the success of their organisations or to their own personal business ethic. Although these two indices locate under different traits of the model (i.e. under 'involvement' and 'consistency' respectively), they are closely related in that according to Denison (2000), the best-performing organisations exhibit excellent team-working skills directed towards shared goals and operating under commonly accepted sets of core values. The importance of these aspects of organisational strength appears to be well understood among respondents within the Hong Kong construction industry and by the HKHD, since 'partnering' workshops and partnered contract arrangements were being launched as a compulsory addition to all contracts

Table 7.2 Descriptive statistics of DOCS cultural indices

Cultural indices	Mean	Std. deviation
Empowerment	3.4217	.3171
Team orientation	3.9482	.2711
Capability development	3.5346	.3027
Core values	3.8790	.3212
Agreement	3.4700	.2536
Coordination and integration	3.2089	.2844
Creating change	3.1662	.3380
Customer focus	3.5664	.3116
Organisational learning	3.4419	.3086
Strategic direction and intent	3.5664	.4701
Goals and objectives	3.4831	.3968
Vision	3.3845	.3081

Source: DOCS individual returns (Coffey, 2005).
Note: 'N' = 23.

issued by the HKHA from 2001 onwards, i.e. coexistent with the period of this research. The standard deviations for 'strategic direction and intent' and 'goal and objectives' were the highest among the indices, meaning that the 157 respondents shared a wider variety of divergence of views on these aspects of their organisation's culture. This is an important result, as these indices are related to the 'mission' trait of organisational culture. According to Denison (2000: 355),

> Perhaps the most important cultural trait of all is a sense of mission... successful organisations have a clear sense of purpose and direction that defines organisational goals and strategic objectives and expresses a vision of what the organisation will look like in the future.

It is a sense of mission that assists organisations to internalise and integrate in order to determine how best to structure themselves and operate in order to reach their desired future state. Committed organisational members who fully understand where a goal-driven organisation is going are more able to contribute to the ultimate success of that organisation. This result also indicates that in many of the companies surveyed, there is some confusion and an unclear view, perhaps even negativity about where the organisation is going, indicating poor communication of goals and strategic objectives within companies. On the basis that the most successful organisations generally had a very strong score for this trait in the DOCS, an assumption is made that the weaker scores of the mediocre and worst-performing companies have contributed most to these high standard deviations observed in the results. The descriptive statistics for the four organisational culture traits and the various PASS constituent scores used in the correlation and linear regression tests are shown in Table 7.3. Although *SPSS version 11.0* (SPSS, Inc., Chicago, ILL.) statistical software also provides the means and standard deviations for the PASS scores of 19 companies (not all of the surveyed companies had PASS results, as has been explained in Chapter 6), these are not considered in relation to the correlation or linear regression tests, as they are not indices which are part of the PASS scoring system, which relies instead on the concept of lower and upper quartiles for the purposes of measuring spread and central tendency. It may be observed from Table 7.3 that of the cultural traits, the mean for 'involvement' was the highest, indicating that a large percentage of the 157 respondents regarded this trait as being strong within their own organisations.

According to Denison (2000), 'involvement' is a trait found in organisations that empower people, encourage team working and develop staff capability at all levels of the organisation. He states that 'People at all levels feel that they have at least some input into decisions that will affect their work and feel that their work is directly connected to the goals of the organisation' (2000: 355). High levels of involvement within companies

Table 7.3 Descriptive statistics of the four DOCS cultural traits and PASS scores

Cultural traits and (PASS scores)	Mean	Std. deviation
Involvement	3.6349	.2550
Consistency	3.5193	.2347
Adaptability	3.3915	.2854
Mission	3.4780	.3754
(Structural and architectural works)	69.1047	1.6452
(Other obligations)	16.7458	1.0599
(Site management)	79.7289	9.9221
(Progress)	71.3405	19.9119

Source: DOCS individual returns (Coffey, 2005).
Note: 'N' = 23 for 'cultural traits' and 19 for PASS scores.

are usually indicative of empowerment, team working and integrated effort to obtain commonly held goals. The four mini-case studies that follow in Chapter 9 of this book also reveal that many of the respondents in the best-performing companies 'highly valued' being delegated high levels of authority (i.e. being empowered) and being part of a team. The standard deviation for 'mission' was the highest among the four traits, meaning that the 157 respondents shared a wider divergence of views on this aspect of their organisation's culture. This derives from the divergence in views on the two indices mentioned previously and supports the view that many of the companies surveyed lack clear strategic direction and intent, or that their managers do not fully comprehend the goals of their organisations. Once again, the lowest scores in this trait were found in the poorest performers measured under PASS.

Factor analysis

Based on the data obtained from the responses to the 60 questions in the questionnaire, a factor analysis was carried out using *SPSS version 11.0* (SPSS, Inc., Chicago, ILL.) followed by the VARIMAX rotation in order to revalidate the DOCS, and Table 7.4 contains the results of the analysis.

Since the Denison Culture Model groups the 60 questions (or variables) into four cultural traits, four factors were specified in the factor analysis and the table is set out to show how the 60 questions group themselves into the specified four factors. The suitability of the dataset for factor analysis was first investigated. In this respect, the sample data are adequate to be explained by the four-factor model as supported by the results from the Bartlett Test of Specificity (with a significance level of 0.0000) and the KMO Measure of Sampling Adequacy at 0.859 (exceeding the threshold of 0.500) (Sharma, 1996; Dulaimi *et al.*, 2002). 'In interpreting factors, a decision must be made regarding which factor loadings are worth

Table 7.4 Results of factor analysis on DOCS using VARIMAX

No	Cultural index	Culture trait	Factors			
			1	2	3	4
1						0.3489
2			0.3640			
3	Empowerment			0.3974		
4					0.4285	
5			0.4253			
6				0.5767		
7				0.5912		
8	Team orientation	Involvement		0.4800		
9				0.4956		
10			0.4172			
11					0.6144	
12				0.4365		
13	Capability development			0.5363		
14				0.5671		
15						0.6125
16				0.5249		
17				0.5629		
18	Core values			0.5664		
19				0.2591		
20			0.4262			
21				0.3576		
22				0.5255		
23	Agreement	Consistency			0.6254	
24						0.5080
25					0.4443	
26						0.5036
27					0.6098	
28	Coordination and integration				0.6126	
29						0.4657
30				0.4327		
31					0.4312	
32			0.4985			
33	Creating change		0.5269			
34					0.5271	
35					0.4183	
36			0.3931			
37			0.4013			
38	Customer focus	Adaptability	0.4238			
39						0.7431
40						0.2630
41			0.4689			

(*Continued*)

Table 7.4 continued

No	Cultural index	Culture trait	Factors			
			1	2	3	4
42			0.5938			
43	Organisational learning					0.4607
44				0.5435		
45				0.4892		
46			0.7034			
47			0.6304			
48	Strategic direction and intent		0.6795			
49			0.7343			
50						0.5345
51	Goals and objectives	Mission	0.5823			
52			0.7121			
53			0.7112			
54			0.6346			
55			0.5512			
56			0.6277			
57			0.6784			
58	Vision					0.5050
59			0.6996			
60			0.3897			
		% Variance	16.192	10.387	8.580	6.750
	Kaiser-Meyer-Olkin (KMO) Measure of Sampling Adequacy		0.859			
		Approx. Chi Square	5051.484			
	Bartlett Test of Spericity	*df*	1770			
		sig	0.000			

Source: DOCS individual returns (Coffey, 2005).

considering.' In a sample of 155 respondents, according to Hair *et al.* (1998), variables with factor loadings lower than 0.45 are considered as insignificant for factor interpretation due to their high standard errors. In this connection, only variables with factor loadings higher than 0.45 are used for factor interpretation in this study. Variables that belong to the cultural traits 'Mission' and 'Adaptability' are grouped in Factor 1 while variables that belong to the cultural traits 'Involvement' and 'Consistency' are grouped in Factor 2. These accounted for 26.58 per cent of the total variance and are regarded as the two most important factors in the factor model. The findings are actually compatible with the original Denison Model, splitting the organisational culture profile into 'internal focus' (that is, derived by 'Involvement' and 'Consistency') and 'external focus' (that is, derived by

'Mission' and 'Adaptability'). This is not really surprising as the original Denison model was partially built on the theory of 'competing values' (Quinn, 1988; Denison and Neale, 1994; Denison and Mishra, 1995), and Denison (1990:15) admits that 'the reconciliation of conflicting demands is the essence of effective organisational culture'.

Qualitative data

Qualitative data and open-ended questions

As mentioned in Chapter 6, in order to triangulate the quantitative data gathered from the 60 questions of the DOCS, two further sets of questions were added to the survey that required an open-ended response.

Self-assessment of performance

The first of these questions required respondents to provide an assessment of their organisation's performance in certain business areas when compared to other companies operating in the same sector. Based on definitions drawn from *InvestorGuide.com* (Murcko, 2004), unless otherwise cited, the specific areas were:

- *Sales/revenue growth*, i.e. the rate at which sales or revenue grew from a previous point of measurement, (e.g. 12-month period to the most recent one).
- *Market share*, i.e. the percentage of the total sales of a given type of product or service that are attributable to a given company.
- *Profitability/ROA* i.e. Return on Assets (ROA) – a measure of a company's profitability equal to a fiscal year's earnings divided by its total assets, expressed as a percentage).
- *Quality of products/services*, i.e. a measure of excellence, usually as viewed by clients or the public at large.
- *New product development*, i.e. a disciplined and defined set of tasks and steps that describe the normal means by which a company repetitively converts embryonic ideas into salable products or services according to the online journal of the Product Development and Management Association (PDMA, 2004).
- *Employee satisfaction*, i.e. although different in every business, the term in this survey is taken to mean – employees with higher job satisfaction:
  - believe that the organisation will be satisfying in the long run;
  - care about the quality of their work;
  - are more committed to the organisation;
  - have higher retention rates;
  - are more productive.

- *Overall organisational performance*, i.e. the results of activities of an organisation or investment over a given period of time.

These latter two definitions are taken from the website of HR consultants Bavendam Research Incorporated, 2003. As may be observed from Table 7.5, over 55.30 per cent of the overall total of returns for all questions were rated 'Average' and this is distributed through the seven sub-questions.

A frequency analysis of all returns to the seven sub-questions, where respondents rated their own company's performance as 'high', 'average' or 'low' compared to others, is shown in stacked bar-graph format in Figure 7.7

Table 7.5 Self-rating question and sub-questions on performance

	Low performer (%)	Average (%)	High performer (%)
Sales/revenue growth	17	63	20
Market share	36	43	21
Profitability/ROA	17	57	26
Quality of products or services	4	48	48
New product development	35	48	17
Employee satisfaction	16	63	21
Overall organisation performance	4	65	31

Source: DOCS individual returns (Coffey, 2005).

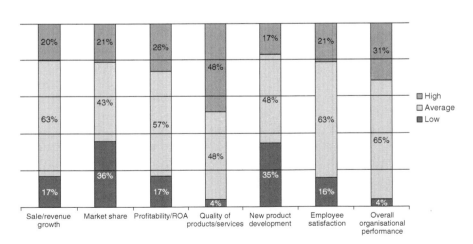

Figure 7.7 Frequency analysis of DOCS performance self-assessment responses.

in order to more clearly depict these results. From this figure, the following general observations can be made:

- Companies rated their 'market share', 'employee satisfaction' and 'new product development' capability as 'low'.
- Only 'quality of product/services' was rated as 'high'.

Open-ended questions on 'value'

The second set of two sub-questions related to the perceptions of individual respondents regarding the concepts of 'value accorded' by their companies to them and 'value achieved' from being a member of their own organisations. As has been mentioned in Chapter 6, these questions were included in order to inform the central core of the Denison Model (i.e. 'beliefs and assumptions') and, in order to place manageable parameters on this aspect of an organisation's culture, the concept of personal and corporate 'values' was examined. This is because as has been investigated in this thesis within the extant literature in Chapter 2 and summed up in the short operational definition adopted for organisational culture definitions taken from Bates and Plog (1990: 7), 'culture is the system of shared beliefs, values, customs, behaviours, and artefacts that the members of society use to cope with their world and with one another, and that are transmitted from generation to generation through learning'. The questions asked were:

1 'What makes me feel most valued?'
2 'What is the thing I most value?'

Because these questions relating to value concepts are open-ended and the responses obtained are in the form of statements varying in length, general comment only is possible here. However, in the set of mini-case studies presented in Chapter 9, the responses to these questions pertaining specifically to the four companies selected to represent different benchmark levels of business success are examined in detail together with their specific DOCS and PASS results, in order to inform the quantitative results of this thesis.

The results obtained to the value-perception questions indicated two sets of major common themes, which may be summarised as follows:

- Money (i.e. salary), position (i.e. promotion prospects) and responsibility (i.e. delegation of authority) were indicative of (1) above to respondents.
- Money (i.e. salary), position (i.e. promotion prospects) and acceptance (i.e. of an individual's ideas followed by capability to implement them) were indicative of (2) above.

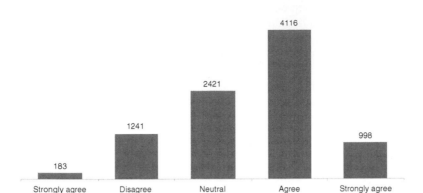

Figure 7.8 Frequency analysis of DOCS ratings distribution.

Consideration of central tendency rating

The possibility of central tendency rating among Asian respondents to questionnaires has already been mentioned in Chapter 6. Ascertainment as to whether or not such a bias existed in the survey returns for the research described in this book therefore became a matter of some interest. A frequency analysis of the raw returns was carried out to ascertain the distribution of ratings for each question based on the five-point Likert-type scale used in the modified DOCS, and the results are as shown in Figure 7.8. As may be observed, there appeared to be no particular strong bias towards central tendency marking; in fact 73 per cent of ratings were outside the 'neutral' box and 57 per cent of ratings were 'agree' or 'strongly agree'. It is clear from this that there was no evidence of any tendency to centrally rate responses obtained from the questionnaires and this was probably due to several possible reasons:

- The nature of the questions in the DOCS is such that it is not generally contentious and supports the giving of answers that can be distinctly positive or negative.
- The DOCS has also been developed very much in a business environment, which means that to the professionals and managers responding to it, the questions are viewed wholly in a business context and therefore can be answered from a genuine business perspective.
- The questionnaires were also designed to be strictly anonymous and therefore responses which were extreme did not pose a threat of external censure or criticism, although it is possible that in some cases, a senior manager or director of a company may have viewed the responses given in the questionnaires, particularly those in paper format.

This result is interesting and significant in that it supports the view of Si and Cullen (1998) and possibly those of Dawes (1972) and Garland (1991) (e.g. standard deviations of responses on certain aspects of the questionnaire) and opposes the views of Spagna (1984), Hofstede and Bond (1988) and Hofstede (1991) mentioned in Chapter 3. It appears also to highlight the need for every researcher to ascertain carefully the number of scale responses to Likert-type questionnaires, based on their expert knowledge of the risk of obtaining central tendency responses from the intended research study population and construct questionnaires accordingly (i.e. to use an even-numbered response scale where central tendency is expected or highly possible).

Chapter summary

This chapter has detailed the specific data and results obtained from examining the Performance Assessment Scoring System (PASS) scores held by the Hong Kong Housing Department (HKHD), and also those derived from the application of the revised Denison Organisational Culture Survey (DOCS) and has then described the findings from these results relevant to the research questions.

Note

1 As it was eventually decided during the course of the study to also use the Premier League Scheme (PLS) matrix to support both the quantitative and qualitative data used to generate the case study element of the research, the higher number of NW2 company returns was actually desirable, as only NW2 companies were represented in the PLS.

8 Detailed statistical analysis of the DOCS and PASS data in relation to the major research questions

Chapter introduction

This chapter provides a quantitative analysis of the data and results described in Chapter 7 but draws no conclusions; nor does it compare these results to those of other researchers already reviewed in Chapters 3 and 4.

Quantitative data: patterns of data for each research question or hypothesis

Data related to the three crucial research questions

Based on the research questions posited in Chapters 1 and 6, the data patterns were investigated using correlation and linear regression tests, and the data, results and findings are summarised in relation to the investigation of the three research questions and nine hypotheses which follow.

Question 1

Do Hong Kong construction companies possessing relatively high combined levels of the four organisational cultural 'traits', i.e. adaptability, involvement, consistency and mission (as indicated by the Denison Organisational Culture Model) perform more successfully on public housing projects than those exhibiting lower levels of those traits?

Question 1 explores the central theoretical concept of this thesis (i.e. a presumed link between organisational culture (OC) and organisational performance (OC)) and so the correlation between the *overall* organisational culture raw scores of contractors and their PASS-measured performance has been investigated in order to test this theory.

The overall organisational culture raw scores were operationalised as the average of the 12 cultural trait scores in 'involvement', 'consistency', 'adaptability' and 'mission' obtained from the responses to the DOCS. PASS-measured performance was generally represented as the overall PASS composite score for the complete years of 2000 and 2001.[1] The results of the Pearson correlation analysis are presented in Table 8.1 and, as can be seen, a positive correlation can be justified between the overall cultural scores and overall PASS scores of the surveyed population at the 5 per cent significance level.

As there are two continuous variables (a dependent variable of performance and an independent variable of culture) and these are not discrete, a regression analysis was carried out (see Table 8.2). Because the hypothesis was stated in its 'null' format and the regression is significant at the 5 per cent level, the null hypothesis is rejected and 'successful' building contractors measured by PASS scores are significantly associated with high combined DOCS 'culture' scores.

As may be seen from Tables 8.1 and 8.2, the H_1 hypothesis (in its null format) can be rejected under correlation and regression testing and therefore a positive answer is justified for Question 1.

Question 2

Are any of the four traits more significant in contributing to success levels than others?

Question 2 really explores the specific concept proposed by Denison (1996, 2000) that each of the four individual cultural traits of 'mission', 'consistency', 'involvement' and 'adaptability' has a strong influence on organisational performance (Denison, 2000), and so the correlation between the four individual cultural trait scores obtained the DOCS and success levels of the surveyed contractors measured as the overall PASS scores were therefore tested first by correlation analysis. The results are shown in Table 8.3.

Table 8.1 Pearson correlation results for OC and OP scores

		Overall cultural score	*Overall PASS score*
Overall cultural score	Pearson correlation	1.000	.532*
	Sig. (2-tailed)		.019
	N	23	19
Overall PASS score	Pearson correlation	.532*	1.000
	Sig. (2-tailed)	.019	
	N	19	23

Source: DOCS individual returns and PASS (Coffey, 2005).

Note: *Correlation is significant at the 0.05 level (2-tailed).

Table 8.2 Regression analysis using PASS as the dependent variable

PASS score	Overall organisational cultural score		Individual organisational cultural trait							
			Adaptability		Involvement		Consistency		Mission	
	R^2	Sig. level	R^2	Sig. level	R^2	Sig. level	R^2	Sig. level	R^2	Sig. level
SW+AW	0.373	*0.006	0.332	0.1	0.323	*0.011	0.313	*0.013	0.316	*0.012
OO	0.301	*0.015	0.268	*0.023	0.148	0.104	0.316	*0.012	0.307	*0.014
SM	0.265	*0.024	0.27	*0.023	0.119	0.148	0.244	*0.032	0.276	*0.021
P	0.256	*0.027	0.276	*0.021	0.116	0.153	0.255	*0.028	0.245	*0.031

Source: DOCS individual returns and PASS scores (Coffey, 2005).

Note: Sig. level * indicates that the regression is at least significant at the 5 per cent level.

Conclusion: The null hypothesis H_1 can be rejected with a risk of less than 5 per cent.

Table 8.3 Pearson correlation results for four individual OC traits and overall PASS scores

		1	2	3	4
1	Overall PASS score				
2	Involvement	.367			
3	Consistency	.522*	.796**		
4	Adaptability	.543*	.795**	.737**	
5	Mission	.526*	.756**	.798**	.820**

Source: DOCS individual returns and PASS (Coffey, and 2003, 2005).

Notes:
**Correlation is significant at the 0.01 level (2-tailed).
*Correlation is significant at the 0.05 level (2-tailed).

The Pearson correlation analysis shows a positive correlation between each of three cultural traits of 'consistency', 'adaptability' and 'mission' with the overall PASS scores at the 5 per cent significance level. Out of these three trait measures of organisational culture strength, 'adaptability' has the most significant correlation, while 'consistency' and 'mission' are of less significance. Only the correlation between 'involvement' and the overall PASS scores cannot be justified in this research study. The null hypothesis can thus be rejected and the research results can therefore justify the answer that out of the four individual cultural traits, those of 'mission', 'consistency', and 'adaptability' do have a significant influence on organisational performance and 'involvement' has no significant influence. In order to provide more robust support for the answer, a regression analysis was undertaken to investigate Hypotheses H_2–H_5 and the results are shown in Table 8.4.

Table 8.4 Regression analysis using PASS as the dependent variable

PASS score	Organisational cultural trait							
	Adaptability (Hypothesis H_2)		Involvement (Hypothesis H_3)		Consistency (Hypothesis H_4)		Mission (Hypothesis H_5)	
	R^2	Sig. level	R^2	Sig. level	R^2	Sig. level	R^2	Sig. level
SW+AW	0.332	0.1	0.323	*0.011	0.313	*0.013	0.316	*0.012
OO	0.268	*0.023	0.148	0.104	0.316	*0.012	0.307	*0.014
SM	0.27	*0.023	0.119	0.148	0.244	*0.032	0.276	*0.021
P	0.276	*0.021	0.116	0.153	0.255	*0.028	0.245	*0.031

Source: DOCS individual returns and PASS scores (Coffey, 2005).

Note: Sig. level * indicates that the regression is at least significant at the 5 per cent level.

Conclusion: The null hypotheses H_2 & H_3 cannot be fully rejected. H_4 & H_5 can be rejected with a risk of less than 5 per cent.

As may be seen, Hypotheses H_2 and H_3 cannot be fully rejected while H_4 and H_5 can be rejected as the regression is significant at the 5 per cent level. Thus the position of 'successful' building contractors on the PASS score league is not significantly associated with 'adaptability' or 'involvement' measured by DOCS. However, the position of 'successful' building contractors on the PASS score league is significantly associated with 'consistency' and 'mission' measured by DOCS. The regression analysis therefore strongly supports similar results to the correlation.

Question 3

Are any combinations of the four traits, based on a horizontal or vertical split of the Denison Organisational Culture Model, more significant in contributing to success levels than others?

Question 3 explores the generic theoretical concept proposed by Denison (1996, 2000) and others (Quinn, 1988; Denison *et al.*, 1995) that there are a set of tensions and contradictions existing within organisations which, depending on whether they are well or badly managed, also have a significant effect on organisational performance. In the Denison Organisational Culture Model, four meaningful combinations of the cultural traits and their directions of influence were established as follows and the theory behind this has been explained in Chapter 5. They are:

- *external focus*: adaptability + mission;
- *internal focus*: involvement + consistency;
- *flexible culture*: adaptability + involvement;
- *stable culture*: mission + consistency.

Measurement of the four combinations was obtained by taking an average of the corresponding constituent cultural scores and investigating the significance of any relationship between them and company success represented by overall PASS score using the Pearson correlation testing methodology. The results of the correlation test are summarised in Table 8.5.

From these results, it may be seen that all four combinations of the cultural traits correlate positively with the overall PASS score at 5 per cent significance level. However, comparatively, 'external focus' and 'stable culture' correlate more significantly than 'internal focus' and 'flexible cultures'.

Once again the multiple regression analysis was run on the same dataset and the results are as shown in Table 8.6. From these results, it may be seen that the null hypotheses H_6, H_8 and H_9 can be rejected whereas H_7 cannot be fully rejected. This means that 'successful' building contractors when measured by PASS do have a significant association with the combined DOCS scores of 'adaptability/mission', 'involvement/adaptability' and 'consistency/mission'. There is no significant association between PASS

Table 8.5 Pearson correlation results for four combined OC traits and direction of influence and overall PASS scores

		1	2	3	4
1	Overall PASS score				
2	External focus	.552*			
3	Internal focus	.463*	.853**		
4	Flexible culture	.486*	.927**	.925**	
5	Stable culture	.543*	.952**	.911**	.866**

Source: DOCS individual returns and PASS (Coffey, 2003, 2005).

Notes:
** Correlation is significant at the 0.01 level (2-tailed).
* Correlation is significant at the 0.05 level (2-tailed).

Table 8.6 Regression analysis using PASS as the dependent variable

PASS score	External (Adaptability + mission) (Hypothesis H_6)		Internal (Involvement + consistency) (Hypothesis H_7)		Flexible (Involvement + adaptability) (Hypothesis H_8)		Stable (Consistency + mission) (Hypothesis H_9)	
Organisational cultural trait	R^2	Sig. level	R^2	Sig. level	R^2	Sig. level	R^2	Sig. level
SW+AW	0.346	*0.008	0.352	*0.007	0.368	*0.006	0.337	*0.009
OO	0.311	*0.013	0.243	*0.032	0.232	*0.037	0.332	*0.01
SM	0.293	*0.017	0.191	0.061	0.214	*0.046	0.282	*0.019
P	0.276	*0.021	0.194	0.059	0.215	*0.046	0.266	*0.024

Source: DOCS individual returns and PASS scores (Coffey, 2005).

Note: Sig. level * indicates that the regression is at least significant at the 5 per cent level.

Conclusion:
1 The null hypotheses H_6, H_8, and H_9 can be rejected with a risk of less than 5 per cent. H_7 cannot be fully rejected.
2 PASS score is mainly dependent on external, stable, and flexible cultural characteristics. Internal cultural characteristics are only significant to the SW+AW and other obligation PASS scores.

scores and 'involvement/consistency'. Put another way, the PASS score is mainly dependent on 'external', 'stable' and 'flexible' cultural characteristics. 'Internal' cultural characteristics are only significant to the SW+AW and other obligation PASS scores.

Regression analysis between cultural scores and PASS scores

Based on the DOCS and PASS data retrieved from the 23 contractors (157) respondents, a regression analysis was carried out using *SPSS version 11.0* (SPSS, Inc., Chicago, ILL.) software to investigate the relationship between the strength of organisational culture of the companies and their 'success', measured by their overall PASS scores. In this analysis, the performance scores measured under the structural, together with architectural works, other obligations, site management and progress aspects, were taken as the dependent variable and regressed one by one. The DOCS scores for the four organisational traits of involvement, consistency, adaptability and mission were taken as the independent variables for single and multiple regression analyses. The regression results summary is shown in Table 8.7.

The regression analysis gives a more robust result than that obtained from the correlation run, and the 'success' of contractors measured by PASS is mainly dependent on the cultural traits of 'consistency', 'adaptability' and 'mission', and 'involvement' is not significant to success apart from to the architectural and structural works scores.

Additional testing of data related to the hypotheses

Based on the DOCS and PASS data of the 23 responding contractors, a correlation between the four DOCS measured cultural traits and the constituent aspects of PASS scores were analysed by using the Pearson correlation testing methodology. The summary results are presented in Table 8.8. From this test, it may be seen that only works scores are significantly associated with 'involvement' whereas all PASS constituent scores are significantly associated with 'consistency', 'adaptability' and 'mission'. The most significant associations between all four DOCS traits are with the most objective of the PASS scores, i.e. the structural (SW) and architectural (AW) works scores. These results differ slightly from the regression results in Chapter 6 where SW and AW were slightly significantly associated with 'involvement', and other obligations, site management and progress were significantly associated with 'adaptability' at the 5 per cent level. The results from both tests are however close enough to strongly support the H_1, H_4 and H_5 hypotheses and some association exists for the H_2 and H_3 hypotheses. Of these 23 responding contractors, 15 had been evaluated for inclusion in the HKHA's Premier League Scheme (PLS) and, as has been discussed in Chapter 6, the success of a contractor can also be reflected somewhat by its position in the PLS, and it was therefore meaningful to investigate the correlation between cultural scores of these 15 contractors and their positions in the PLS. Table 8.9 shows the results.

The results from this test are interesting in that when compared to the earlier regression results and the Pearson correlation test between the OC traits and overall PASS scores shown in Chapter 7, there is some difference in the outcomes. Generally, 'successful' contractors drawn by reference with the PLS rankings have a significant association with the 'consistency', 'mission', 'internal focus' and 'stable culture' dimensions of DOCS. As far as the association with 'mission' and 'consistency' is concerned, this tallies exactly with the regression results obtained when overall PASS scores and DOCS traits are tested; however, the DOCS dimension of 'internal focus'

Table 8.7 Summary of regression analysis of individual cultural traits using PASS as the dependent variable

| PASS score | Overall multiple regression | | Significance of individual cultural trait in multiple regression | | | | | | | |
| | R² | Sig. level | Adaptability (Hypothesis H_2) | | Involvement (Hypothesis H_3) | | Consistency (Hypothesis H_4) | | Mission (Hypothesis H_5) | |
			Coeffnt	Sig. level	Coeffnt	Sig. level	Coeffnt	Sig. level	Coeffnt	Sig. level
SW+AW	0.38	0.129	1.523	0.551	1.168	0.635	1.253	0.69	0.05	0.982
OO	0.381	0.128	1.06	0.52	−1.481	0.355	2.234	0.278	0.386	0.788
SM	0.344	0.178	13.839	0.386	−14.631	0.343	14.458	0.461	4.188	0.762
P	0.361	0.154	37.441	0.242	−32.57	0.289	42.319	0.282	−4.313	0.875

Source: DOCS individual returns and PASS scores (Coffey, 2005).

Notes: Conclusions by inspection of regression coefficient values:
1 PASS score is mainly dependent on consistency, adaptability and mission. However, mission has negative influence to progress score in the multiple regression model.
2 Involvement is not significant to PASS score except for SW+AW score.
3 In general, the results of multiple regressions are in line with that of individual regression analyses.

Table 8.8 Pearson correlational analysis of DOCS cultural indices and constituent PASS aspect scores

		1	2	3	4	5	6	7
1	Involvement							
2	Consistency	.796**						
3	Adaptability	.795**	.737**					
4	Mission	.756**	.798**	.820**				
5	SW & AW scores	.568**	.560*	.577**	.562*			
6	OO scores	.385	.562*	.518*	.554*	.504*		
7	Site management	.345	.493*	.520*	.526*	.594**	.904**	
8	Progress	.341	.505*	.525*	.495*	.668**	.862**	.946**

Source: DOCS individual returns (Coffey, 2003, 2005).

Notes:
** Correlation is significant at the 0.01 level (2-tailed).
* Correlation is significant at the 0.05 level (2-tailed).

(involvement/consistency) was not significantly associated with overall PASS scores in regression testing. 'Stable' orientation (consistency/mission) was significantly associated in both regression and correlation testing with overall PASS scores and rankings on the PLS. The reason for this is possibly that only the top six companies in PLS are also the best scoring companies under PASS (the other companies vary in their league positions between the two leagues). In addition, many of the criteria measured under the PLS are of a more subjective nature, closer to the non-works scores of PASS which also do not appear to correlate or regress as strongly with DOCS as the more objective measurements of success. This will be discussed holistically in Chapter 10.

Results related to each hypothesis

Based on the research hypotheses (stated below in the 'null' premise) posited in Chapters 1 and 3 using the Denison Organisational Culture Survey (DOCS) and the Hong Kong Housing Department (HKHD)'s Performance Assessment Scoring System (PASS), the data patterns related to each of the research hypotheses were as follows.

*   With reference to *Research Question 1*:

    H_1: Position of 'successful' building contractors on the PASS score league is not significantly associated with high (i.e. third and fourth quartile) combined organisational culture scores, as measured by the Denison Organisational Culture Model.

    The results show that in companies possessing combined DOCS scores falling in the *third/fourth* quartile of the Denison OC Model, their 'strong' organisational culture is positively and significantly associated

Table 8.9 Pearson correlational analysis of OC and PLS ranking scores

		Premier league score	Involvement	Consistency	Adaptability	Mission	External focus culture	Internal focus culture	Flexible culture	Stable culture	Overall cultural score
Premier league score	Pearson correlation	1,000	.414	.588*	.298	.564*	.464	.522*	.370	.593*	.502
	sig. (2-tailed)		.125	.021	.281	.028	.081	.046	.175	.020	.057
	N	15	15	15	15	15	15	15	15	15	15
Involvement	Pearson correlation	.414	1,000	.796**	.795**	.756**	.809**	.952**	.941**	.811**	.903**
	sig. (2-tailed)	.125		.000	.000	.000	.000	.000	.000	.000	.000
	N	15	23	23	23	23	23	23	23	23	23
Consistency	Pearson correlation	.588*	.796**	1,000	.737**	.798**	.808**	.943**	.807**	.921**	.898**
	sig. (2-tailed)	.021	.000		.000	.000	.000	.000	.000	.000	.000
	N	15	23	23	23	23	23	23	23	23	23
Adaptability	Pearson correlation	.298	.795**	.737**	1,000	.820**	.940**	.810**	.953**	.829**	.918**
	sig.(2-tailed)	.281	.000	.000		.000	.000	.000	.000	.000	.000
	N	15	23	23	23	23	23	23	23	23	23
Mission	Pearson correlation	.564*	.756**	.798**	.820**	1,000	.966**	.819**	.833**	.970**	.938**
	sig. (2-tailed)	.028	.000	.000	.000		.000	.000	.000	.000	.000
	N	15	23	23	23	23	23	23	23	23	23
External focus culture	Pearson correlation	.464	.809**	.808**	.940**	.966**	1,000	.853**	.927**	.952**	.973**
	sig. (2-tailed)	.081	.000	.000	.000	.000		.000	.000	.000	.000
	N	15	23	23	23	23	23	23	23	23	23
Internal focus culture	Pearson correlation	.522*	.952**	.943**	.810**	.819**	.853**	1,000	.925**	.911**	.950**
	sig. (2-tailed)	.046	.000	.000	.000	.000	.000		.000	.000	.000
	N	15	23	23	23	23	23	23	23	23	23
Flexible culture	Pearson correlation	.370	.941**	.807**	.953**	.833**	.927**	.925**	1,000	.866**	.961**
	sig. (2-tailed)	.175	.000	.000	.000	.000	.000	.000		.000	.000
	N	15	23	23	23	23	23	23	23	23	23
Stable culture	Pearson correlation	.593*	.811**	.921**	.829**	.970**	.952**	.911**	.866**	1,000	.970**
	sig. (2-tailed)	.020	.000	.000	.000	.000	.000	.000	.000		.000
	N	15	23	23	23	23	23	23	23	23	23
Overall cultural score	Pearson correlation	.502	.903**	.898**	.918**	.938**	.973**	.950**	.961**	.970**	1,000
	sig. (2-tailed)	.057	.000	.000	.000	.000	.000	.000	.000	.000	
	N	15	23	23	23	23	23	23	23	23	23

Source: DOCS individual returns and PASS scores (Coffey, 2005).

Notes:
* Correlation is significant at the 0.05 level (2-tailed).
** Correlation is significant at the 0.01 level (2-tailed).

with high levels of organisational performance measured by PASS. There appears to be no significant association between higher combined DOCS scores and a high level of overall organisational performance measured by a high-ranking position in the PLS (where companies are actually in the PLS), i.e. the PLS companies do not have higher 'success' levels based on their combined DOCS scores. However, the 'null' hypothesis can be rejected based on the tests undertaken.

- With reference to *Research Question 2*:

H_2: Position of 'successful' building contractors on the PASS score league is not significantly associated with 'adaptability' in their organisational culture as measured by the Denison Organisational Culture Model.

The results show that in companies possessing a DOCS 'adaptability' score falling in the *third/fourth* quartile of the Denison OC Model, this aspect of their 'strong' organisational culture is positively correlated and highly significantly associated with higher PASS scores in 'architectural and structural works'. However, although a high 'adaptability' score is also positively correlated to higher PASS scores in 'other obligations', 'site management' and 'progress', the association is less significant. There appears to be no significant association between 'adaptability' and a high level of overall organisational performance measured by a high-ranking position in the PLS (where companies are actually in the PLS), the PLS companies do not have higher 'success' levels based on their DOCS 'adaptability' scores. The 'null' hypothesis cannot therefore be fully rejected.

H_3: Position of 'successful' building contractors on the PASS score league is not significantly associated with 'involvement' in their organisational culture as measured by the Denison Organisational Culture Model.

The results show that 'involvement' correlates positively with 'architectural and structural works', but not to the other aspects of PASS performance. Hypothesis H_3 cannot therefore be fully justified or refuted and neither can the alternative hypothesis be fully supported. There appears to be no significant association between 'involvement' and a high level of overall organisational performance measured by a high-ranking position in the PLS (where companies are actually in the PLS); the PLS companies do not have higher 'success' levels based on their DOCS 'involvement' scores. The 'null' hypothesis cannot therefore be fully rejected.

H_4: Position of 'successful' building contractors on the PASS score league is not significantly associated with the 'consistency' measure in their organisational culture as measured by the Denison Organisational Culture Model.

The results show that in companies possessing a DOCS 'consistency' score falling in the *third/fourth* quartile of the Denison OC Model, this aspect of their 'strong' organisational culture is positively correlated and highly significantly associated with higher PASS scores in 'architectural and structural works' and 'other obligations'. However, although a high 'consistency' score is also positively correlated to higher PASS scores in 'site management' and 'progress', the association is less significant. There appears to be a positive association between 'consistency' and a high level of overall organisational performance measured by a high-ranking position in the PLS (where companies are actually in the PLS), i.e. the PLS companies do have higher 'success' levels based on their DOCS 'consistency' scores. The 'null' hypothesis can therefore be rejected.

H_5: Position of 'successful' building contractors on the PASS score league is not significantly associated with the perception of 'mission' in their organisational culture as measured by the Denison Organisational Culture Model.

The results show that in companies possessing a DOCS 'mission' score falling in the *third/fourth* quartile of the Denison OC Model, this aspect of their 'strong' organisational culture is positively correlated and highly significantly associated with higher PASS scores in 'architectural and structural works' and 'other obligations'. However, although strong culture is also positively correlated to higher PASS scores in 'site management' and 'progress', the association is less significant. There is also a positive association between 'mission' and a high level of overall organisational performance measured by a high-ranking position in the PLS (where companies are actually in the PLS), i.e. the PLS companies do have higher 'success' levels based on their DOCS 'mission' scores. The 'null' hypothesis can therefore be rejected.

- With reference to *Research Question 3* using the Denison Organisational Culture Model:

H_6: Position of 'successful' building contractors on the PASS score league, is not significantly associated with third and fourth quartile 'Denison' rankings in 'adaptability/mission'.

The results show that in companies possessing a DOCS combined 'adaptability/mission' score falling in the *third/fourth* quartile of the Denison OC Model, their 'strong' external focus represented by these aspects of OC is positively associated with high PASS scores in 'architectural' and 'structural works', 'other obligations', 'site management' and 'progress'. There no strong positive association with a high level of overall organisational performance measured by a high-ranking position in the PLS (where companies are in the PLS). The 'null' hypothesis can therefore be rejected only in relation to PASS score success measurements.

H₇: Position of 'successful' building contractors on the PASS score league is not significantly associated with third and fourth quartile 'Denison' rankings in 'involvement/consistency'.

The results show that in companies possessing a DOCS combined 'involvement/consistency' score falling in the *third/fourth* quartile of the Denison OC Model, their 'strong' internal focus is positively associated with high PASS scores only in 'architectural' and 'structural works' and 'other obligations'. There is also a positive association with a high level of overall organisational performance measured by a high-ranking position in the PLS (where companies are in the PLS); however, the 'null' hypothesis cannot therefore be fully rejected.

H₈: Position of 'successful' building contractors on the PASS score league is not significantly associated with third and fourth quartile 'Denison' rankings in 'involvement/adaptability'.

The results show that in companies possessing a DOCS combined 'involvement/adaptability' score falling in the *third/fourth* quartile of the Denison OC Model, their 'strong' management flexibility is positively associated with high PASS scores in 'architectural' and 'structural works', 'other obligations', 'site management' and 'progress'. There is also a positive association with a high level of overall organisational performance measured by a high-ranking position in the PLS (where companies are in the PLS). The 'null' hypothesis can therefore be rejected.

H₉: Position of 'successful' building contractors on the PASS score league is not significantly associated with third and fourth quartile 'Denison' rankings in 'consistency/mission'.

The results show that in companies possessing a DOCS combined 'consistency/mission' score falling in the *third/fourth* quartile of the Denison OC Model, their 'strong' management flexibility is positively associated with high PASS scores in 'architectural' and 'structural works', 'other obligations', 'site management' and 'progress'. There is also a positive association with a high level of overall organisational performance measured by a high-ranking position in the PLS (where companies are in the PLS). The 'null' hypothesis can therefore be rejected.

Chapter summary

Chapter 7 described in detail the data and results drawn from the conducting of the research for this thesis, and this chapter has then analysed those data and recorded the basic findings. However, the reader will note that no conclusions are drawn in this chapter and neither are the results compared to those of other researchers already reviewed in earlier chapters. The methodologies followed in collecting the data were described in detail in

Chapter 6, as were the quantitative and qualitative analysis techniques applied to the data.

The conclusions resulting from the findings described in this chapter will be discussed in the context of the literature in Chapter 10, together with the implications for theory and further research.

Note

1 This selected PASS score time period is in line with the time period of conducting the survey of companies using the DOCS.

9 Four Hong Kong construction mini-case studies

Chapter introduction

Chapter 8 described and summarised the results derived from the administration of the DOCS and also outlined the results based on secondary data drawn from PASS. This chapter presents mini-case studies taken from four companies off the Hong Kong Housing Authority (HKHA)'s List of Building Contractors representing different benchmark levels of business success and based on comparative positions measured under PASS and the Premier League Scheme (mentioned in both Chapters 1 and 6).

The mini-case studies

In order to further inform the quantitative results of this thesis, an examination in depth on a mini-case study basis was carried out on the qualitative datasets from four companies as follows:

- Mini-case Study No. 1 (Company 9) – the *second ranked* company on the PASS score league and one of the six PLS contractors.
- Mini-case Study No. 2 (Company 18) – the *first-ranked* company on the PASS score league and one of the six PLS contractors.
- Mini-case Study No. 3 (Company 22) – a *middle-ranked* (but rapidly improving company) under both the PASS score league and the PLS.
- Mini-case Study No. 4 (Company 12) – the *lowest-ranked* company on the PASS score league and not included in the PLS, although also an NW2 group company on the HKHA list.

The annual reports of three of these companies selected for the mini-case studies were obtained and the documents were examined in detail for content specifically related to organisational culture, values, beliefs, mission and effectiveness. In the case of Company 12, no annual reports were available and so at an interview held with the quality manager, various simplified datasets were collected (including simple turnover and profits, quality and environmental policies). The content of the open-ended question answers relating to value perceptions of respondents from these companies, as well as the results of the self-assessment of performance under DOCS, were also examined in order to further inform the results presented in Chapter 7. The content of all of the latter text was extracted into a simple database and analysed for patterns and matches using simplified content analysis software called *Concordance Version 3.20* (Watt, 2004). This facilitated a revelation of some relationships between the corporate philosophies of the three better performers and the perceptions of their staff on performance and concepts of value, but in relation to the poorer performer a reverse picture was observed, which tended to support more strongly the results obtained from the quantitative area of study in the thesis.

Mini-case Study No. 1 (Company 9)

Business background and objectives

The group of companies of which Company 9 is a major member had been achieving sustainable financial performance during the 10-year period starting around 1990. With their engagement on some of the most of the sizeable construction contracts in Hong Kong and projects in other Asian countries, the group has attained an annual average value of work completed of approximately HK\$5 to 6 billion. Figure 9.1 shows the profile of work over the period leading up to the research study being carried out.

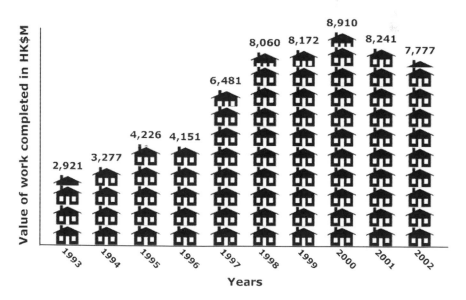

Figure 9.1 Company 9: turnover (1993–2002).
Source: Coffey (2003).

Although the group has several specific construction sectors in which it operates, government and HKHA projects accounted for almost 25 per cent of its total construction turnover in 2001/2002. This company's turnover was in fact lowered in both 2001 and 2002 due to the 'drying-up' of large government capital works contracts (in particular former Airport Core Projects), the curtailment and downsizing of the HKHA's substantial housing programme which has been ongoing since 2000 and the general deterioration of private sector residential capital projects following the Asian financial crisis. This phenomenon may also be seen in the two mini-case studies that follow (i.e. Companies 18 and 22). The three main sources of the group's construction projects are the government and related institutional development projects, a major Hong Kong commercial property development company's projects and development projects from other private property developers. The total contracts-on-hand of the group as at 31 August 2002 were approximately HK$20 billion as shown in Figure 9.2.

Corporate mission and values

Company 9 was founded in 1960 and was restructured in 1973 to become an affiliate of a large Hong Kong development company, allowing greater diversity of its construction operations and the ability to be involved in private real estate development. A further restructuring of the parent group in 1993 saw Company 9 become the principal member of a new holding company which as a group offered project planning and management,

Figure 9.2 Company 9: sources of work (31 August 2002).
Source: Coffey (2003).

building and civil construction and electrical and mechanical (E & M) services, design-and-build procurement and building materials and E & M equipment supplies. Based on the company's 2004 Annual Report, their stated vision, mission, values and 'company' slogan were as follows:

Vision
To serve our clients in the construction industry, turning conception into reality, emphasising innovation, safety and quality.

Mission: We build to our clients' satisfaction.
We care for the growth and well-being of our people.
We help build and shape a better living environment.

Values: Highest standard of integrity
Harmonious team work and employee loyalty
Efficiency and effectiveness in operation
Commitment to quality, safety and environmental protection

Slogan: "We deliver and we care"

The company's webpage of the same year (2004) featured *A Letter from Management* which attributes the company's success to the following organisational *culture*:

> [The Company] owes its success and its ability to generate higher profits to its shareholders to an effective blend of audacity and caution, anticipation and adaptation, competitiveness and creativity, and a workforce of the highest calibre dedicated to the creation of a better, cleaner and more prosperous life for the community.

In order to ascertain whether the values, mission and perception of company strength was aligned throughout the company down to the front-line managerial and contract staff levels, the results of the open-ended survey questions that follow are examined in some depth.

Perception of 'values' by staff of Company 9

There were seven questionnaire responses (of which *four* were usable in respect of this specific exercise) from this company, and as may be seen from the 'age', 'service', 'qualifications' and 'post or department', most of the respondents mirrored the 'typical profile' of respondents developed in

Table 9.1 Company 9 DOCS open-ended question responses (value perceptions)

Age	Service	Quals	Post or department	What makes me feel most valued?	What is the thing I most value?
40–49	>15	M Deg	Gen Mgt	Company continues to grow and sustains profitable turnover	Systematic changes introduced by me are being adopted and implemented without sacrificing company's reputation and profitability
30–39	10–15	M Deg	Const Mgt	Money along with job satisfaction	Job satisfaction together with raise and promotion
40–49	>15	B Deg	Const Mgt	Trust offered by senior management	Job satisfaction
40–49	>15	M Deg	Purchasing	No response	No response
50–59	>15	Pg Dip	Gen Mgt	No response	No response
40–49	6–10	M Deg	HR Dept	No response	No response
40–49	10–15	M Deg	Admin Dept	Encouragement from boss	Job satisfaction and respect from company

Source: Coffey (2003).
Notes:
M Deg = Master's degree.
B Deg = Bachelor's degree.
Pg Dip = Postgraduate diploma.

Chapter 6, i.e. 30 to 49 years of age, holding a technical diploma/degree, middle or senior or construction manager, 6 to 15 years with the company. The results of the response to the two open-ended questions on "what makes me feel most valued" and "what is the thing I most value" are shown in Table 9.1.

Self-rating of performance aspects by staff of Company 9

The results drawn from the self-rating matrix within the DOCS in the seven responses (of which *all* were usable in respect of this specific exercise) from this company are shown in Table 9.2.

DOCS and PASS results

An examination of the organisational culture profile of Company 9 shown in Figure 9.3 indicates that the three 'strongest' component indices are 'core values', 'team orientation' and 'strategic direction and intent', and the two lowest scoring are 'coordination and integration' and 'creating change'.

A closer look at the individual trait level and the actual responses within the items of those trait groups reveals that the following specific questions received the highest/lowest scores, as shown in Table 9.3.

Table 9.2 Company 9 DOCS responses (performance self-assessment)

Age	Service	Quals	Post or department	1	2	3	4	5	6	7
40–49	>15	M Deg	Gen Mgt	H	H	H	H	L	A	H
30–39	10–15	M Deg	Const Mgt	H	H	H	H	L	A	A
40–49	>15	B Deg	Const Mgt	H	H	H	H	H	H	H
40–49	>15	M Deg	Purchasing	A	A	L	A	A	L	A
50–59	>15	Pg Dip	Gen Mgt	H	H	H	H	H	A	H
40–49	6–10	M Deg	HR Dept	A	H	H	H	L	A	H
40–49	10–15	M Deg	Admin Dept	H	H	H	A	A	L	A
			Average rating	85.72	92.86	85.72	85.72	42.86	42.86	78.57

Source: Coffey (2003).

Notes:
1 = Sales/Revenue Growth.
2 = Market Share.
3 = Profitability/ROA.
4 = Quality of Product/Services.
5 = New Product Development.
6 = Employee Satisfaction.
7 = Overall Organisational Performance.

H = High performance (100 pts).
A = Average performance (50 pts).
L = Low performance (0 pts).

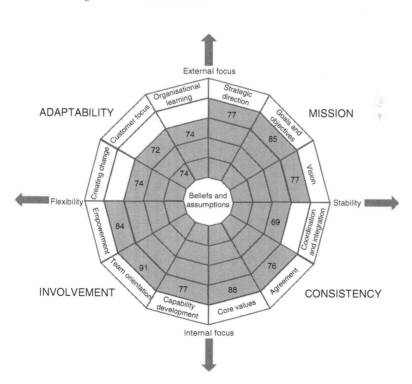

Figure 9.3 Company 9: DOCS profile.

Table 9.3 Company 9 highest/lowest line item scores

Highest scores

☑ Company competitive advantage over others depends on the capability of its staff.
☑ Team work, rather than hierarchy, helps to get task completed.
☑ Teams are essential elements of our work.
☑ Managers and leaders should have consistency between what they speak and practise.
☑ Company has its own main culture.

Lowest scores

☒ We rarely encounter obstacles when we try to change things.
☒ Our short-term thinking seldom undermines our long-term vision.
☒ It is usually quite easy to reach consensus on key issues.
☒ It is easy to coordinate projects among different departments.

Source: Modified Denison Organisational Culture Survey (Coffey, 2005).

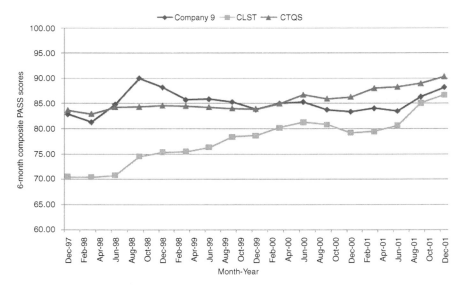

Figure 9.4 Company 9: PASS score trend.

The PASS score trend of Company 9 is shown in Figure 9.4 and indicates that the company has managed to track fairly consistently the Composite Target Quality Score (CTQS) throughout the period of performance under consideration. However, it is noticeable that around the same time that the company's overall workload began to drop, there was a corresponding dropping and levelling of performance, and this was reversed towards the end of the measurement period.

Observations and discussion

These results indicate that whilst the respondents felt that their team-work capabilities contributed to the company's success, managers did not appear to 'walk the talk' and thus the strategic direction of the company sometimes suffered. All senior managers consulted had a strong sense of the company's core values of quality and customer service and none felt that they could do better in any other company amongst their competitors. It appears that the perceived inability to form a consensus may be directed at the top management level and the middle managers seem to feel that the vision of those above them is somewhat short term. This was borne out during focus group interviews with respondents and during discussions statements such as "we have good ideas for how the company can become even better but there is no system to give this view to the bosses". Another construction manager complained that "management don't always realise how we protect their backs for them, we never get any credit for our efforts as a group, only complaints when our PASS scores drop to low. At the same time that we began to have some performance deficiencies, wages were cut by 25 per cent because of dropping workload and the recession in Hong Kong." Respondents had a good grasp of how successful they were in terms of constructing public sector housing and their responses to the 'value' questions indicated a concentration on more intrinsic rewards such as trust, adoption of ideas and delegation of authority, and these responses are supportive of the strong elements measured in the company by DOCS, thus confirming the reliability and validity of the model.

HKHA perceptions

This company is a Group NW2 contractor in the HKHA List of Building Contractors (New Works) and was first admitted to the list in 1999 (relatively recently compared to most other contractors on the HA lists who entered in 1989 or 1990, or who were already at the time of list promulgation in long-established business relationships with the HKHA). Since admission, the company has undertaken five contracts to date since listing, of which they had completed one contract and had two contracts in hand at the time the research was undertaken. This company was the *second-ranked* company on the PASS score league and one of the six PLS contractors.

Mini-case Study No. 2 (Company 18)

Business background and objectives

According to their corporate website,

> The parent group [of Company 18] was founded in 1971 ... and is a diversified group engaged in property development, construction and construction materials with interests in Hong Kong, the Chinese

Mainland and North America. The Group's corporate culture and long term objectives are based on its commitment to quality, innovation and excellence. In 1997 the construction and construction materials businesses were grouped under one parent, which was listed on the Hong Kong Stock Exchange in February of the same year. Today, the latter group's core businesses include quality housing development in the Chinese Mainland, cement, construction, construction materials and global materials trading.

The construction group's well-established corporate culture and longer term corporate objectives, as set out in its corporate mission statement below, are based on a well-proven focus on quality and excellence throughout the group. Company 18 was the first contractor in Hong Kong to achieve ISO 9002 certification for building works in 1992 and was also accredited with ISO 14001 environmental management system certification and Occupational Health and Safety Series (OHSAS) 18001 certification in 1999 and 2000, respectively. The company's subsidiary operations have won various prestigious awards and recognition from the media and clients, such as the HKHA's "Contractor of the Year Award" for its outstanding achievements in construction quality, project progress and site safety management. In safety and health management, Company 18 has also won "Best Building Works Contractor Awards" from the HKHA for seven consecutive years since 1992, achieving more than 40 gold, silver and bronze awards in addition to a number of other awards such as the "Safety Training Award" and the "Innovative Safety Device Award". Company 18 has also been awarded the ETWB's "Considerate Contractors Award" for three consecutive years since 1996 and in 2000.

Company 18 may be said to have specialised in public housing and institutional buildings in Hong Kong. A smaller sister company is also involved in interior fitting out and building renovation works for the HKHA. The company was selected by the HKHA in 2000 as one of its six Premier League Scheme (PLS) contractors. Figure 9.5 shows the financial results of the parent group for the 10 years leading up to and including the time when this research study was undertaken.

This company's results, as in those of Company 9 described earlier and Company 22 that follow, indicate a pattern of steadily rising financial growth in terms of turnover for the period from 1994 onward. This rising trend peaked in the 1999/2000 financial year and began to decline in the following two years.

Corporate mission and values

The corporate mission statement and company objectives for the different facets and focuses of the organisation are represented below and have been paraphrased and adjusted to preserve the anonymity of Company 18.

CORPORATE MISSION STATEMENT

The achievements of an organisation are the result of the combined efforts of all the individuals in the organisation working towards common objectives. It is therefore very important that everyone ... is committed to these objectives to ensure their accomplishment.

As a business investment, we have to provide attractive financial returns to our shareholders. Profitability is essential since it is the prime measure of our corporate success and the source of our growth and development. In order to achieve sustained profitability, we must aim at customer loyalty and long-term revenue generation by providing quality products and services at competitive prices.

To provide quality products and services with consistency, we have to ensure that our staff is highly competent, motivated, and well rewarded for their efforts. In addition, we must strive to provide an environment in which our staff can excel and develop under a set of shared beliefs and a firmly established corporate culture. The long-term success of [the Company] is based on the following corporate value and objectives:

Profit: to achieve sufficient profit to provide an attractive return to our shareholders and to finance our growth.

Customers: to provide our clients with quality services and products.

Staff: to provide an environment whereby our people can excel, develop and grow with the company.

Management philosophy: to provide an environment that encourages and rewards merit and team effort.

Corporate culture: to cultivate a set of shared beliefs on which all our policies and actions are based.

Perception of 'values' by staff of Company 18

There were 10 questionnaire responses (of which 9 were usable in respect of this specific exercise) from this company. The results of the response to the two open-ended questions on "what makes me feel most valued" and "what is the thing I most value" are shown in Table 9.4.

As may be observed when comparing the company's value perceptions with those of the respondents, there is good alignment between both. The company espouses good team work, performance, high-quality product and

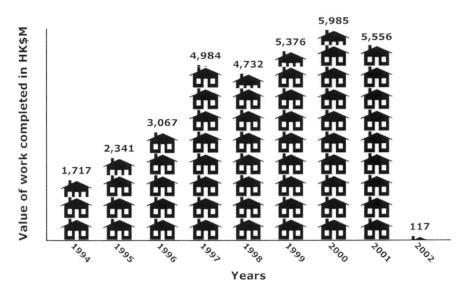

Figure 9.5 Company 18: turnover (1994–2002).

profitability, and managers responding to the open-ended questions on value perceptions closely parallel very similar espoused values of achievement, success, team working, trust and loyalty, and recognition of efforts. The history of this organisation explains this close alignment as the establishment of a strong and well-knit 'quality' culture always being the main aspiration of the CEO from the foundation of the company in 1971. The majority of respondents have more than 15 years', service with the company which provides generous benefits for its staff; managers are sometimes given shares in the company and performance bonuses, and workers are allocated above-average health and welfare benefits compared to other, similar companies.

Self-rating of performance aspects by staff of Company 18

The results drawn from the self-rating matrix within the DOCS in the 10 responses from this company are shown in Table 9.5.

In terms of 'profitability/ROA' and 'overall organisational performance' most respondents have rated the company's achievement level comparatively high, whereas 'new product development' carries a comparatively low rating. All ratings when compared to those of the other companies investigated by the mini-case studies are high, indicating a very positive internal view of the company's overall success within the sector.

Table 9.4 Company 18 DOCS open-ended question responses (value perceptions)

Age	Service	Quals	Post or department	What makes me feel most valued?	What is the thing I most value?
30–39	10–15	Pg Dip	HR Dept	Team work and cooperation	Team work and cooperation
40–49	>15	B Deg	Gen Mgt	Achieving goals and being respected	Having a team of loyal staff with common goals; achieving results
40–49	>15	B Deg	Const Mgt	Trust	Team work
40–49	>15	B Deg	Estimating	Culture of company towards perfection	Working in my company
40–49	>15	B Deg	Const Mgt	Our company's people	Good performance of the company
40–49	>15	B Deg	Eng Dept	Position	Job satisfaction
40–49	>15	B Deg	Eng Dept	Contribution being recognised	Working relationship between colleagues
50–59	>15	B Deg	Gen Mgt	The corporate mission statement of the company which sets out the values and belief of the company	A team of talented and determined staff that always strives for high performance standards
40–49	>15	M Deg	Gen Mgt	No response	No response
30–39	10–15	B Deg	Const Mgt	Achievement gained in successfully completing the assignments given by my company	Success of company and individual

Source: Coffey (2003).

Notes:
M Deg = Master's degree.
B Deg = Bachelor's degree.
Pg Dip = Postgraduate diploma.

DOCS and PASS results

Company 18 had the 'strongest' overall DOCS score and also possessed the 'highest overall performance' score measured under PASS for the period being considered in the research study. The overall DOCS profile, highest and lowest line item scores and related PASS score trend graph are as shown in Table 9.6, and Figures 9.6 and 9.7. In relation to their detailed PASS scores, this company consistently achieved high 'input' and 'output' scores, meaning that the company's high-quality workmanship on a large number of construction projects being undertaken during the period of performance being considered was supported by strong and productive management input to these projects. From the DOCS profile, the three 'strongest' cultural traits are 'core values', 'team orientation' and 'goals and objectives', and the three lowest scoring are 'coordination and integration', 'customer focus' and 'organisational learning'.

Table 9.5 Company 18 DOCS responses (performance self-assessment)

Age	Service	Quals	Post or department	1	2	3	4	5	6	7
30–39	10–15	Pg Dip	HR Dept	A	H	H	H	A	H	H
40–49	>15	B Deg	Gen Mgt	A	A	H	A	A	H	H
40–49	>15	B Deg	Const Mgt	A	H	A	H	A	H	H
40–49	>15	B Deg	Est Dept	H	H	H	H	H	H	H
40–49	>15	B Deg	Const Mgt	H	H	H	A	A	A	A
40–49	>15	B Deg	Eng Dept	H	A	H	A	H	A	A
40–49	>15	B Deg	Eng Dept	H	A	H	A	H	A	H
50–59	>15	B Deg	Gen Mgt	H	H	H	H	H	H	H
40–49	>15	M Deg	Gen Mgt	H	H	H	H	H	H	H
30–39	10–15	B Deg	Const Mgt	A	H	H	H	A	H	H
			Average rating	80.00	85.00	95.00	80.00	75.00	85.00	90.00

Source: Coffey (2003).

Notes:
1 = Sales/revenue growth.
2 = Market share.
3 = Profitability/ROA.
4 = Quality of product/services.
5 = New product development.
6 = Employee satisfaction.
7 = Overall organisational performance.

H = High performance (100 pts).
A = Average performance (50 pts).
L = Low performance (0 pts).

Table 9.6 Company 18 highest/lowest line item scores

Highest scores
☑ Teams are essential elements of our work.
☑ Company has its own main culture.
☑ I am clear on the company's strategic direction.
☑ Leaders have a long-term viewpoint.
☑ Company has a clear strategy for the future.

Lowest scores
☒ We can easily reach consensus on different issues, no matter how difficult they are.
☒ Staff of different departments have the same points of view.
☒ We rarely encounter obstacles when we attempt to change things.
☒ The ways we do business are flexible and easy to change.
☒ Our short-term thinking seldom underlines the long-term vision.

Source: Modified Denison Organisational Culture Survey (Coffey, 2005).

A closer look at the individual trait level and the actual responses within the items of those trait groups reveals that the following specific questions received the highest/lowest scores, as shown in Table 9.6.

The PASS score trend of Company 18 is shown in Figure 9.7 and indicates that the company has consistently tracked the Composite Target Quality Score (CTQS) (and sometimes set or exceeded this threshold) throughout the period of performance under consideration. However, it

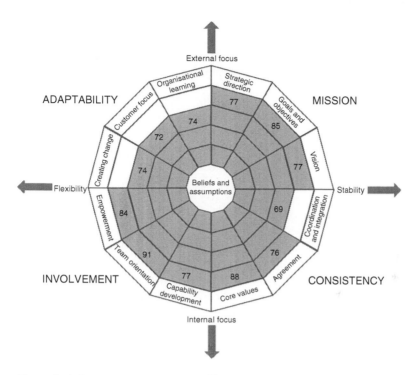

Figure 9.6 Company 18: DOCS profile.

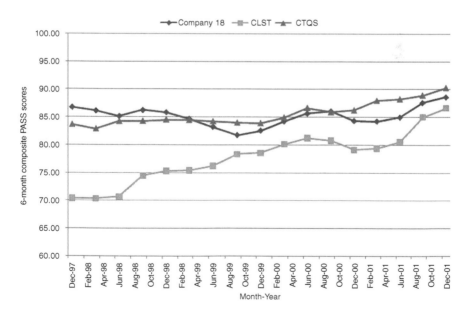

Figure 9.7 Company 18: PASS score trend.

is noticeable that as in the previous case study, around the same time as the company's overall workload began to drop, there was a corresponding (but comparatively minor) dropping and levelling of performance. This was reversed towards the end of the measurement period.

Observations and discussion

These results, together with discussions with respondents at the focus group session, indicate that all respondents had a deep feeling for the company's strong culture, long-term vision of the directors, strategic leadership and direction (in other words a sense that the company has a 'future') and perception that everyone was working as a team. Although the low scores were relatively high compared to the respondents from other companies, there appeared to be a prevalent view that departments were fairly rigid and sometimes formed obstructions to consensus, and that thinking was sometimes too short term when compared to the view of the directors. During the focus group session held with respondents, some interesting comments arose when the question 'What obstacles occur internally between departments of the company?' was asked. Most managers hastily reinforced the view that the company was very integrated and everyone 'pulled together for the team'. One manager (known to be more outspoken than the rest of the group) volunteered that 'our estimating department who sets the tender prices rarely take advice from front-line managers on pricing of subcontract work ... their prices are much too low for the quality of work desired by the client to be attained'. This opened up the views of others who shared an opinion that 'we are all company men and understand the company's and customer's requirements ... we have specialised our knowledge of PASS to ensure we consistently meet standards ... but our subcontractors are outside the system and often undermine the team working situation'. Most of the respondents agreed that the major barrier to successful quality on contracts was due to varying competencies and motivation of domestic subcontractors who are difficult to manage. The fact that the three strongest traits measured for Company 18 under DOCS were 'mission', 'consistency' and 'involvement' was interesting, as it did not accord with the results of statistical analysis of the comparison of PASS results and DOCS for the whole population. In the case of the latter, these results indicated that for the whole population, 'adaptability' had the strongest correlation with PASS scores, although 'consistency' and 'mission' were also correlated but less significantly. The reasons for this may be due to the fact that the combined traits of 'mission' and 'consistency' are signs of a stable culture, whereas the combined traits of 'involvement' and 'consistency' represent an 'internal focus' of an organisation. Company 18 has been such a long-term partner of HKHD over the years that there may well be a feeling amongst front-line managers that 'they know better than the client' and therefore do not have to be so customer focused. In addition, the fact that they have also had long-term success in terms of PASS scores and tender awards does not actually require them to be so adaptable, as there is little need for them to significantly alter their cultural profile unless performance were to deteriorate or they cease to win tenders. Denison (2000) observes in relation to such apparent contradictions that organisations which are stable, well integrated and have great strategic orientation often find it hard to focus adequately on the customer but the best organisations manage to resolve these contradictions.

HKHA perceptions

This company is a Group NW2 contractor in the Housing Authority Building (New Works) list and was first admitted to the list in 1989. Since admission, the company has undertaken over 20 contracts and had eight contracts in hand at the time the research was undertaken. The company has generally had an excellent performance record from being first listed, which has included gaining many 'best contractor' and 'best project' awards, and it was also elected on to the PLS in 2002 at the top position. In all ways, this company may be said to be the 'most successful' business partner based on measures operated by the HKHD.

Mini-case Study No. 3 (Company 22)

Business background and objectives

Company 22, established in 1958, is the construction arm of a parent company that was listed on the Hong Kong stock exchange in 1991. Its major businesses include building construction, maintenance, fitting out, plumbing and drainage contracting, building service engineering consultancy, sale and manufacture of building materials and pre-cast products, property development, and research on developing high-tech building technology. The group employs over 1,500 people in Hong Kong and Mainland China. Figure 9.8 shows the results of the group for the 10 years leading up to and including the time when this research study was undertaken.

This demonstrates clearly the improvement in the group's turnover, which had begun to grow steadily from 1994 and was sustained until 2000. As mentioned in the previous cases, there was a sharp drop in the years 2000/2001/2002. However, despite this downturn, during the year 2000 (i.e. at the time the research for this thesis was being undertaken), Company 22 held some 25 per cent of the HKHA's new works building contracts, amongst a group comprising some 22 other building firms.

Corporate mission and values

Since 1986, Company 22 has built more than 30 public housing estates for the HKHA, providing accommodation for more than 300,000 people, and the company's corporate webpage and a contemporary Annual Report of

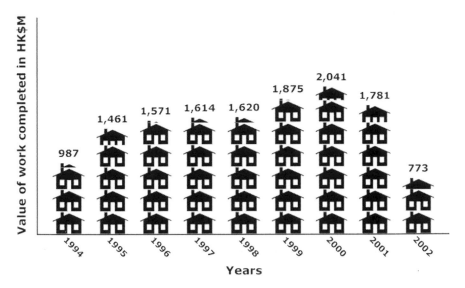

Figure 9.8 Company 22: turnover (1994–2002).

the group at the time the survey was undertaken (2001/2002) states the following company business objectives:

COMPANY OBJECTIVES

- To show the greatest respect to the client, and deliver quality constructions and services to satisfy the needs and expectations of the clients and society so as to increase our market competitiveness.
- To establish a fair and long-term relationship with clients, subcontractors, suppliers and employees.
- To ensure construction work is carried out in a high-quality and efficient manner through proper support to subcontractors and suppliers.
- To develop staff potentials by providing a suitable working environment and career development opportunities.
- To raise the Group's efficiency and quality of works through the adoption of ISO 9000 Quality System.
- To make continuous improvement to the operations of the Group and encourage staff to do so constantly.

Perception of 'values' by staff of Company 22

There were eight questionnaire responses (of which six were usable in respect of this specific exercise) from this company, and the results of the response

Table 9.7 Company 22 DOCS open-ended question responses (value perceptions)

Age	Service	Quals	Post or department	What makes me feel most valued?	What is the thing I most value?
30–39	10–15	M Deg	Gen Mgt	To satisfy customers and develop new products	Good direction forward in the company
40–49	>15	Pg Dip	Const Mgt	To complete projects on time without any complaint from the client	To receive a good reputation from other departments and the boss
50–59	6-10	M Deg	Gen Mgt	Dependence of the company on my existence	Ability to make important decisions and implement them
40–49	6–10	Pg Dip	Const Mgt	When what I do can make the most profitable benefits for the company	When my opinion can be accepted by the top management of the company
40–49	10–15	M Deg	Gen Mgt	No response	No response
40–49	10–15	M Deg	Engineering	Top management acceptance of my work and efforts	High degree of freedom in my working methods
40–49	>15	Pg Dip	Const Mgt	No response	No response
40–49	>15	Pg Dip	Site Agent	Quality of products and services	Honesty

Source: Coffey (2003).
Notes:
M Deg = Master's degree.
B Deg = Bachelor's degree.
Pg Dip = Postgraduate diploma.

to the two open-ended questions on "what makes me feel most valued" and "what is the thing I most value" are shown in Table 9.7.

It may be observed from these responses that similar to the situations described for Companies 9 and 18 in the two preceding mini-case studies, but perhaps not as strongly demonstrated, there appears to be alignment between some of the corporate and individual value perceptions, particularly in the common desire to provide quality service and products and to enhance company reputation. There is however a noticeably much stronger expression of a need for acceptance and recognition, together with greater delegation of authority and autonomy, and this is indicative of a less rigidly managed company. During the period of the study, greater management autonomy was handed to the senior and middle managers, and it is perhaps for this reason that performance improvements were made.

Self-rating of performance aspects by staff of Company 22

The results drawn from the self-rating matrix within the DOCS in the 10 responses from this company are shown in Table 9.8.

These results are interesting as, other than in the areas of 'quality of product/services' and 'new product development', the opinion of managers on the company's other aspects of performance is comparatively low. In particular, 'sales/revenue growth' and 'market share' are considered to be poor. This appears to point to a general view amongst managers that despite the company performing well in terms of its quality output and its ability to innovate, its market share does not match its efforts in terms of overall rewards. The company, possibly realising this general feeling amongst its managers, subsequently devoted a section on its website espousing what it wishes to achieve by way of performance and the benefits it anticipated such performance should provide to the company. The following extract is included to highlight this set of aspirations:

- [The Group] believes that Quality, Safety and Environmental Management are the keys to maintain the Group's competitiveness in difficult times. Therefore the Group has never compromised in these areas despite the fall in tender prices. In fact, effective quality management has resulted in quality workmanship hence reduced costs for remedial works.

  [Author's note: The Company obtained its ISO 9002 (i.e. quality management system) certification in early 1993 and its ISO 14000 (i.e. environmental management system) certification in 2001, which made it one of the first companies in the construction industry to be certified to both of these international standards in Hong Kong].

- Moreover, performance in these respects will enhance the Group's chance of tendering for future government contracts. The Group's efforts were also evidenced by the winning of several government and HKHA awards by Company 22 during the year.

  [Author's note: The Company has over the last 10 years or so won numerous awards from the HKHA and other government bodies in the areas of quality, safety, innovative construction, considerate contractor, productivity and best overall performance].

- [The Group's] ongoing research and development program improves quality of work, minimizes costs and increases efficiency. The group's 'Semi-Precast System' was one of the many significant advances in construction techniques. This system is specially designed to meet HKHA's requirements for the construction of pre-cast façade panels used in the housing blocks.

  [Author's note: Company 22 was awarded a prestigious quality award in respect of its pre-cast products by the Trade and Industry Department of Hong Kong in October 2000].

Table 9.8 Company 22 DOCS responses (performance self-assessment)

Age	Service	Quals	Post or department	1	2	3	4	5	6	7
30–39	10–15	M Deg	Gen Mgt	H	A	H	H	H	H	H
40–49	>15	Pg Dip	Const Mgt	A	L	L	A	H	A	A
50–59	6–10	M Deg	Gen Mgt	A	A	A	H	H	A	A
40–49	6–10	Pg Dip	Const Mgt	L	L	A	H	H	A	H
40–49	10–15	M Deg	Gen Mgt	A	A	A	A	H	A	A
40–49	10–15	M Deg	Engineering	L	L	A	H	H	A	A
40–49	>15	Pg Dip	Const Mgt	A	A	A	H	H	H	H
40–49	>15	Pg Dip	Site Agent	A	A	A	H	H	H	H
			Average rating	43.75	31.25	50.00	87.50	100.00	68.75	75.00

Source: Coffey (2003).

Notes:
1 = Sales/revenue growth.
2 = Market share.
3 = Profitability/ROA.
4 = Quality of product/services.
5 = New product development.
6 = Employee satisfaction.
7 = Overall organisational performance.

H = High performance (100 pts).
A = Average performance (50 pts).
L = Low performance (0 pts).

DOCS and PASS performance

Company 22 had a relatively 'strong' overall DOCS score and its 'overall performance' score measured under PASS for the period being considered in the research study shows improvement in the latest quarterly periods. The overall DOCS profile and related overall PASS scores are shown in Figures 9.9 and 9.10. In relation to their detailed PASS scores, this company consistently achieved moderate 'input' and medium to high 'output' scores, meaning that the company's quality of workmanship on a large number of construction projects being undertaken during the period being considered was supported by an increasing level of effective management input to these projects. From the DOCS profile, the three 'strongest' cultural traits are 'core values', 'goals and objectives' and 'team orientation', and the three lowest scoring are 'coordination and integration', 'agreement' and 'customer focus'.

A closer look at the individual trait level and the actual responses within the items of those trait groups reveals that the specific questions in the DOCS instrument received the highest/lowest scores, as shown in Table 9.9.

The PASS score trend of Company 22 is shown in Figure 9.10 and indicates that the company commenced the measurement period with an average level of performance half-way between the CTQS and the Composite Lower Score Threshold (CTQS) benchmarks. This is very indicative of Company 22 in the late 1990s since, due to their long-standing relationship with HKHD, they produced reasonable quality construction output on projects, but rarely excelled at that time.

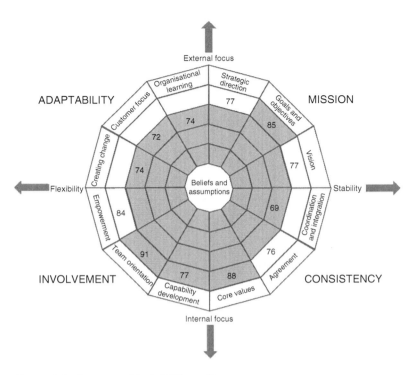

Figure 9.9 Company 22: DOCS profile.

The scores fell sharply at the end of 1998 due to some isolated quality problems that had a devastating effect on PASS scores on two significant projects at a time when the company also had only a few projects on site under construction. However, following a significant reorganisation within the company and a thorough review and action on its subcontracting partners, the quality levels began to rise steadily from early 1999 onward. From mid-1999 to the end of 2001, the company's performance rose steadily until they had become a significant performer on the PASS score league. Unlike the situation of the previous case studies, there was no noticeable dropping of scores as workload downturned, and it may also be observed that the drop in workload was arrested and increased in the last year as shown in the financial table. Companies 9, 18 and 22 were part of the reason that overall quality levels of the list of contractors improved in the latter half of the research period, as they collectively possessed a significantly large percentage of the HKHA's projects during the period.

Observations and discussion

These results indicate an extremely high overall culture score in the 'involvement' trait (two fourth quartile indices) as well as third quartile

Table 9.9 Company 22 highest/lowest line item scores

Highest scores

☑ Cooperation between different departments is highly recommended by the company.
☑ Company will continuously invest in its staff's techniques.
☑ Company competitive advantage over others depends on the capability of its staff.
☑ Company will never ignore the interests of the customers when decisions are made.
☑ Company has long-term goal and directions.
☑ I am clear on the company's strategic direction.

Lowest scores

☒ Staff of different departments have the same points of view.
☒ It is easy to coordinate projects among different departments.
☒ There is a clear agreement to decide the right and wrong way to work.
☒ Problems seldom arise because we have the techniques required to carry out the work properly.
☒ We rarely encounter obstacles when we attempt to change things.
☒ Our short-term thinking seldom undermines the long-term vision.

Source: Modified Denison Organisational Culture Survey (Coffey, 2005).

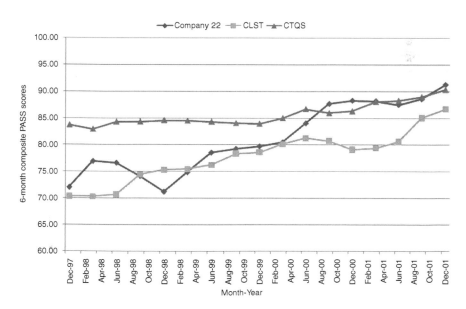

Figure 9.10 Company 22: PASS score trend.

scores for all other traits and indices. The highest line-item scores indicate a strong belief amongst managers that the 'company' will provide for them in terms of training and strategic motivation. During the focus group session with respondents, many strongly expressed their appreciation for the company's attention to their training needs and also to extending their capabilities in terms of encouraging them to take on further qualifications and form quality groups to discuss issues on PASS and contracts generally. One older manager did however indicate that because the company had different business interests (i.e. construction, building materials and precast products and IT), it had become much larger as an organisation and it was not always so easy to coordinate with other departments as had been the case back in the late 1990s. Other managers disagreed and felt that the team working in the company was strong, and the more diverse activities had helped the company avoid financial difficulties once the current recession had 'bitten' the construction industry. Rather more difficult to analyse were the lower scored line items as some of these conflicted with the highest scoring, i.e. whilst it was obvious that most respondents valued the company's investment in training, the reverse-coded item "Problems seldom arise because we have the techniques required to carry out the work properly" was marked low. This was discussed in the focus group session, and the view of the group was that some of the managers felt strongly that the training was more directed at the site levels of staff and not at the managerial level. Most respondents said that if they were given time off, they would want to undertake more training, perhaps even for a Master's degree, but this was unlikely to happen as their workload was too heavy. The other interesting conflicting response was that whilst the survey indicated that the company's long-term goals and directions were clearly spelled out and known to all managers, the short term view necessary to deal with 'fire-fighting' and changing client standards and demands sometimes undermined this strategic overview. One other widely held view was that the PASS had been a major driver in the company to improve quality, and the close attention to the details of scores and the assessments on site had been an invaluable tool to the front-line site supervisors, i.e. the PASS had been a change driver in the company. This may explain why, although 'adaptability' was not the highest score, it was across all the indices in the third quartile which is relatively high amongst the firms surveyed.

HKHA perceptions

This company is a Group NW2 contractor in the Housing Authority Building (New Works) list and was first admitted to the list in 1989. Since admission, the company has undertaken over 40 contracts but at the time the research was undertaken it had only four or six contracts in hand.

It is a *middle-ranked* (but rapidly improving company) under both the PASS score league and the PLS. Subsequent to the research being carried out, the company was successful in being admitted to the PLS in 2003.

Mini-case Study No. 4 (Company 12)

Business background and objectives

Company 12 was established in 1978 and was one of the earliest companies to work for the HKHD. The company's financial results from 1993 to 2002 are shown in Figure 9.11. As may be seen from these results, in financial terms this company has not performed nearly as well as the other three case study companies and, since the firm was suspended from tendering in 2000 and did not have many major jobs for other clients, public or private, due to 'adverse' reports on other government projects, at the time of writing they are virtually inactive in the construction sector.

Corporate mission and values

Company 12 appears to have no corporate statement of mission and/or values other than in its formal quality and environmental policies. As the latter is modelled to a large extent on the former, it is sufficient to set out the statements relating to quality (essentially the variable being measured by PASS) and the quality management system or organisational

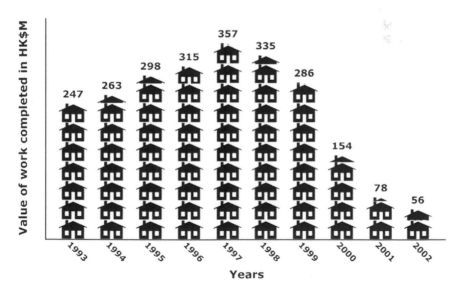

Figure 9.11 Company 12: turnover (1993–2002).

fabric (essentially the variables being measured by DOCS) which are as follows:

QUALITY POLICY

- [The Company] is committed to a quality management system which is appropriate to the purpose of the organisation and complies with the requirements of the ISO 9000 series standards, including the requirement to continually improve the effectiveness of the quality management system.
- Quality objectives are established and reviewed.
- This policy is communicated and understood by all people within the organisation and it is regularly reviewed for continuing suitability.

The company only recently developed a website and at the time of the research did not appear to publish any formal annual reports (although under the tax laws of Hong Kong applicable to all companies operating under a business registration, sole proprietorship or certificate of incorporation, annual profits tax returns are completed together with audited accounts). This made it very difficult to obtain meaningful data on company perceptions of value. It is clear from the company's long association with HKHA that it values its ongoing and long-standing public sector workload and, providing its capacity for consecutive contracts has not been exceeded, it has at times managed to perform well, as evidenced by some of the PASS scores that follow (i.e. late 1998) on at least one project for the client. Awarding too many projects to the company appears to cause problems, as management resources become too 'stretched' and there is also a reliance on subcontractors who are unfamiliar with the quality requirements of the HKHD. Further comment on this may be found in the section on Company 12's DOCS and PASS scores which follows.

Perception of 'values' by staff of Company 12

There were four questionnaire responses (of which only two were usable in respect of this specific exercise) from this company. The results of the response to the two open-ended questions on "what makes me feel most valued" and "what is the thing I most value" are shown in Table 9.10. Responses received are pitched at a fairly fundamental level of expectation, particularly when one considers that they have been drawn from senior managers in the company. Team working and continued employment appear to satisfy the value perceptions of one manager and acceptance of suggestions and adequate salary the other. It is difficult to make further comment on

Table 9.10 Company 12 DOCS open-ended question responses (value perceptions)

Age	Service	Quals	Post or department	What makes me feel most valued?	What is the thing I most value?
40–49	6–10	B Deg	Purchasing	No response	No response
30–39	4–6	Pg Dip	Site Agent	No response	No response
30–39	6–10	Pg Dip	Const Mgt	Cooperation from colleagues	Hard work and responsibility
30–39	6–10	M Deg	Engineering	My suggestions are adopted	Salary

Source: Coffey (2003).

Notes:
M Deg = Master's degree.
B Deg = Bachelor's degree.
Pg Dip = Postgraduate diploma.

Table 9.11 Company 12 DOCS responses (performance self-assessment)

Age	Service	Quals	Post or department	1	2	3	4	5	6	7
40–49	6–10	B Deg	Purchasing	A	L	A	A	L	L	A
30–39	4–6	Pg Dip	Site Agent	–	L	–	A	L	L	–
30–39	6–10	Pg Dip	Const Mgt	–	L	–	–	–	A	A
30–39	6–10	M Deg	Engineering	L	L	A	A	L	A	A
			Average rating	25.00	0.00	50.00	50.00	0.00	25.00	50.00

Source: Coffey (2003).

Notes:
1 = Sales/revenue growth.
2 = Market share.
3 = Profitability/ROA.
4 = Quality of product/services.
5 = New product development.
6 = Employee satisfaction.
7 = Overall organisational performance.

H = High performance (100 pts).
A = Average performance (50 pts).
L = Low performance (0 pts).

this as there is little high-level corporate mission or objective information to compare against, and this may be a reason for the continuous poor performance of the company in the period under study.

Self-rating of performance aspects by staff of Company 12

The results drawn from the self-rating matrix within the DOCS in the four responses (of which *all* were somewhat usable in respect of this specific exercise) from this company are shown in Table 9.11. Not unexpectedly, the majority of responses indicated a comparatively low self-rating in all aspects of performance. This aside, there was some indication that respondents felt their 'quality of product/services' and 'overall organisational performance' were at least average, which is in fact a fairly accurate perception of their position in terms of PASS scores and the score leagues operated by HKHD.

DOCS and PASS performance

Company 12 had a relatively 'weak' overall DOCS score, and its 'overall performance' score measured under PASS, apart from the three quarterly periods from September 1998 to June 1999, was well below the Composite Lower Score Threshold (CLST) for the period being considered in the research study. The overall DOCS profile, highest and lowest cultural indices line items and related overall PASS scores are shown in Figures 9.12 and 9.13, and Table 9.12. In relation to their detailed PASS scores, this company consistently attracted poor 'input' and less than satisfactory 'output' scores, meaning that the company's quality of workmanship on only one or two construction projects being undertaken during the period being considered was the worst amongst all new works contractors, attracting a substantial number of adverse reports which led to their suspension from further tendering in mid-2000. From the DOCS profile shown in Figure 9.12, the two 'strongest' cultural indices are 'customer focus' and 'team orientation',

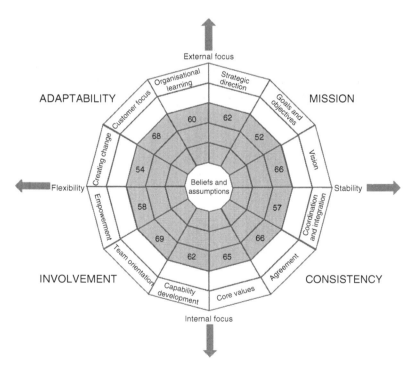

Figure 9.12 Company 12: DOCS profile.

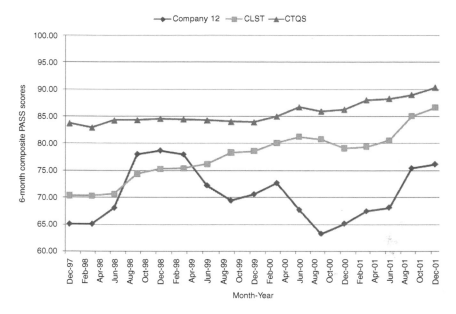

Figure 9.13 Company 12: PASS score trend.

and 'vision' and 'agreement' are also just mid-third quartile. The two lowest scoring indices are 'goals and objectives' and 'creating change'.

An examination of the scores of responses reveals that the specific questions in DOCS received the highest/lowest scores as shown in Table 9.12.

The PASS score trend of Company 12 is shown in Figure 9.13 and indicates that the company has consistently achieved PASS scores well below the CLST (falling below this threshold means that the company will draw 'adverse' reports which will prevent it from tendering for HKHD projects and will also curtail opportunities to work for other government agencies). This is indicative of the mediocre performance of Company 12 since the late 1990s, as may be observed from its score trend which has gone in a diametrically opposite direction to the overall trend of companies on this chart, and in fact has fallen at a more severe rate than the CLST and CTQS have risen! Falling scores and adverse reports trigger a vicious cycle for companies since, if they cannot tender for work their cash flow suffers, and this makes it difficult to hire capable subcontractors or staff, or or to obtain bank guarantees required by government clients to cover the capital requirements of their listed companies.

Observations

The company refused to let its respondents participate in a focus group session, although the quality manager accepted an invitation to an interview

Table 9.12 Company 12 highest/lowest line item scores

Highest scores

- ☑ Cooperation between different departments is highly recommended by the company.
- ☑ Teams are essential elements of our work.
- ☑ Company competitive advantage over others depends on the capability of its staff.
- ☑ Managers and leaders should have consistency between what they speak and practice.
- ☑ Troubles arise when workers ignore the core values.
- ☑ Customer involvement will directly influence the company's decisions.

Lowest scores

- ☒ Faced with changes by competitors to business environment, company can have a good response.
- ☒ Company can continuously adopt the latest and improved ways of work.
- ☒ We rarely encounter obstacles when we attempt to change things.
- ☒ Company encourages innovations and risk taking and gives rewards if these are seen.
- ☒ I am clear on the company's strategic direction.
- ☒ Our short-term thinking seldom undermines the long-term vision.

Source: Modified Denison Organisational Culture Survey (Coffey, 2005).

following the completion of the survey questionnaire by respondents (of which he was one). He confirmed what appeared obvious from the survey responses and the answers to the open-ended questions on values and also the self-rating questions, i.e. that the company had no drive and staff had little motivation. They were in fear for their job security, knowing only too well that the company had huge cash flow difficulties. The QM noted that morale had in fact been much better when the company was actively tendering, and noted that on one complex job for HKHD from the period September 1998 until mid-1999, the company had achieved good enough scores to be awarded a tendering opportunity but had been unable to take advantage of this due to downsizing and lack of competent subcontractors. The low culture scores for 'goals and objectives' and 'coordination and integration' as well as 'creating change' are particularly telling, as these are dimensions of culture that determine the ability to motivate staff, ensure that different departments and teams work well together, and allow adaptability to change to meet different environments and client requirements. It is probably the inability to change and adapt that has been this company's major downfall, and certainly supports the reliability and validity of the DOCS in terms of showing a significant relationship between weak organisational culture and poor performance.

HKHA perceptions

This company at the time that the research was undertaken was a Group NW2 contractor in the Housing Authority Building (New Works) list and was first admitted to the list in 1989. Since admission, the company has undertaken several medium-sized contracts although it had already been suspended at the time the research was being undertaken, which could be the reason for its poor survey return rate, although the company was at least prepared to give some returns, and that is indicative of its interest in obtaining feedback in relation to its major client's projects despite difficulties. The company has generally had a poor performance record from being first listed and was not considered for the PLS, being the lowest scored on that league when all companies were assessed for consideration of PLS listing. Therefore, based on utilising measures operated by both the HKHA and HKHD, Company 12 was the *lowest-ranked* company on the PASS score league and was not included in the PLS.

Chapter summary

This chapter presented mini-case studies taken from four companies on the HKHA's List of Building Contractors and each represented different benchmark levels of business success based on comparative positions measured under PASS, and, where applicable, the Premier League Scheme (PLS). In order to further inform the quantitative results of this thesis, qualitative data obtained from these four companies were examined and related to their detailed OC traits under their DOCS profiles and their PASS individual aspect scores. The general findings from these mini-case studies related to how the information contained in them fits into the overall results of the research described previously are discussed in Chapters 7 and 8, and conclusions regarding their significance and contribution to this thesis and theory in general are expounded in Chapter 10.

10 Conclusions and future research directions

Chapter introduction

We now come to the end of our journey looking at the often complex and certainly significant relationships between organisational culture and company performance and effectiveness, in the latter chapters rooted firmly in the construction industry arena in Hong Kong. Despite the specific location and environment of the research described previously, the general conclusions now discussed can equally well be considered in relation to construction companies universally, although exact replication of the research in other countries could be more challenging owing to the specfiic nature of the HKHD PASS used to measure the variable of organisational performance. That stated, many other organisations in a wide variety of countries do choose to use many of the performance and effectiveness metrics described in Chapter 4, and DOCS has since the time of this author's research begun to find its way into, and is being used successfully in many other countries (e.g. Russia, the Netherlands, China, Indonesia). The sections that follow give an overview of the research results covered in Chapters 7 and 8 and also propose new areas of research or ways in which the author's own research could be usefully extended to add further to the extant knowledge on the overall subject matter of this book. Hopefully by the end of this chapter the reader with have a greater understanding of the field of organisational culture research, and most importantly an 'understanding of organisational culture in the construction industry'.

Qualitative data

An analytical discussion of the qualitative data from the research

In order to inform further the quantitative results of this thesis, an examination and in-depth analysis of the results obtained from the four mini-case studies described in Chapter 9 was carried out and the general findings from these results are described in this chapter. For the benefit of the reader, the four companies investigated in the mini-case studies were as follows:

- Company 9 – the second-ranked company on the PASS score league and one of the six PLS contractors.
- Company 18 – the first-ranked company on the PASS score league and one of the six PLS contractors.
- Company 22 – a middle-ranked (but rapidly improving) company under both the PASS score league and the PLS.
- Company 12 – the lowest-ranked company on the PASS score league and not included in the PLS, although also an NW2 group company on the HKHA list.

Specific findings from this qualitative examination of the corporate and individual value and performance perceptions of the four companies were expounded in Chapter 7 and generally there appeared to be a parallel understanding of company goals and objectives by individuals within the two best-performing organisations, but there was less alignment in the improving company and it was difficult to examine this relationship at all in the worst-performing case. Self-perceptions of performance and company standing in the two best-performing companies were close to their actual standing based on their PASS scores and also their position in the PLS. In the case of the improving company, only some of the self-perceptions of respondents accurately reflected their true position. In the case of the worst performer, there was a closer relationship between the self-perceptions of respondents on their performance and status.

General findings related to answers to open-ended questions on 'value perceptions'

The overall responses to open-ended questions on internal and external 'values' perceptions were evaluated and any common patterns investigated and analysed. Based on an overall comparison of the corporate mission and values statements and the responses to the two DOCS sub-questions, (1) 'What makes me feel most valued?' and (2) 'What is the thing I most value?' The following general findings were made.

The results obtained to the value-perception questions indicated two sets of major common themes, which may be summarised as follows:

- Money (i.e. salary), position (i.e. promotion prospects) and responsibility (i.e. delegation of authority) were indicative of (1) to respondents.

- Money (i.e. salary), position (i.e. promotion prospects) and acceptance (i.e. of an individual's ideas followed by capability to implement them) were indicative indicators of (2).

The common value 'driver' in both responses is reward for effort and the recognition resulting from that reward. Deeper values of 'trust' and 'empowerment', as well as 'respect' for knowledge and 'acceptance' to allow ideas to be realised are evident in these results. If companies can realise the importance that these deeper drivers play in terms of motivating staff and ensure that they do not rely only on financial reward alone to push up performance, individuals can be brought together to form strong and efficient teams that provide a foundation for the building of strong cultures and can demonstrate effective, efficient and class-winning performance.

General findings related to answers to self-rated question on 'performance'

The results obtained from the self-perception questions relating to performance and company status in the market indicated some significant overall response trends, which may be summarised as follows:

- respondents rated their 'market share', 'employee satisfaction' and 'new product development' capability as 'low';
- only 'quality of product/services' was rated as 'high'.

This is an interesting but not unsurprising finding, since the views of employees at middle-management or lower levels of the company are not always focused on strategic issues that some of these self-perceptions are designed to measure. It appears from these results that the company value objectives are not always shared and espoused in the same way by the directorate as they are by the general and construction management levels in many of the companies surveyed. The CEOs, executive and associate directors (somewhat understandably, since they are also often shareholders in the mid-sized to larger companies) valued most highly company-related success factors, while the construction managers and administrative and general management staff value more pragmatic factors such as money, job satisfaction, trust and encouragement.

Major research findings from the analysis of results

The overriding reason for carrying out the research described in this book was originally in fulfilment of the author's own university doctoral programme requirements. However, interest in the whole area of organisational studies emanated from an interest that originated in the author's own working field of over 20 years (i.e. the construction industry), in particular in the question of why some construction companies performed well consistently and others only sporadically, or not at all. What is different about the best performers and what gives rise to their success? This interest extended to closely observing the different management styles in action and the kind of organisational 'feel' that different companies possessed, and this triggered the author to start to read the extant literature about organisational and other cultural studies. Reviewing the literature in great detail soon gave rise to the various discussions on the difference between organisational culture and climate, the arguments about which research methodology was best to study the concept of organisational culture and the dialogues that abound on the way to measure organisational effectiveness. Throughout the literature, I kept coming across the various studies that had gone before on the perceived link between organisational culture and performance, and one name cropped up with regularity: Dr Daniel Denison. I began my studies by reading his book entitled *Corporate Culture and Organisational Effectiveness*, and from that point on my research idea began to develop.

In terms of the success of this thesis to answer my own basic query above, I believe that my studies have borne fruit in that I have a fairly deep understanding of what it is within organisations that makes a successful or unsuccessful company. Indeed, the results reported by Denison and others have been closely replicated in my study, thus giving further support to the premise that there is a relationship between 'strong' organisational cultures and effective business performance, that the Denison Organisational Culture Survey is useful in other ethnic and organisational cultures from those in which it was first developed and that there are other ways to measure effectiveness rather than the previous financial or the more recent score-card models.

Conclusions about each major research question

Based on the research questions stated in Chapters 1 and 3, the conclusions relating to each were as follows.

With reference to Research Question 1

Do Hong Kong construction companies possessing relatively high combined levels of the four organisational cultural 'traits' (i.e. adaptability, involvement, consistency and mission) (as indicated by the Denison Organisational Culture Model) perform more successfully on public housing projects than those exhibiting lower levels of those traits?

Based on the Pearson and regression analyses described in Chapter 8, it is evident that companies possessing relatively high combined levels of the four organisational cultural 'traits' (i.e. 'adaptability, involvement, consistency and mission') (as indicated by the Denison Organisational Culture Model) *do* perform more successfully on public housing projects than those exhibiting lower levels of those traits. These findings support the view that a relationship does exist between organisational culture and effectiveness/performance (and also that companies possessing 'strong' organisational cultures perform better than those exhibiting 'weaker' OC profiles. The results also show that in companies possessing high combined DOCS scores, their 'strong' organisational culture is positively and significantly associated with high levels of organisational performance measured by PASS. There was no significant association between higher combined DOCS scores and a high level of overall organisational performance measured by a high-ranking position in the PLS (where companies are actually in the PLS). While these results demonstrate the existence of the link between organisational culture and performance and also show that this link exists whether the measures of effectiveness used are financial (Denison, 1990) or non-financial (as in the 1997 version of PASS), it does not necessarily negate the PLS system as a measure of business effectiveness. The latter is not an absolute comparative measure such as PASS. It is a more generic measure establishing a threshold above which companies are considered to possess overall attributes which make them suitable partners to handle HKHA's most complex or prestigious projects, i.e. it is a measure of how the corporate profiles of companies satisfy certain general benchmark criteria.

With reference to Research Question 2

Are any of the four traits more significant in contributing to success levels than others?

The Pearson correlation analysis described in Chapter 8 shows a positive correlation between each of three cultural traits of 'consistency', 'adaptability' and 'mission' with the overall PASS scores at the 0.05 significance level. Out of these three trait measures of organisational culture strength, 'adaptability' has the most significant correlation, while 'consistency' and 'mission' are of less significance. Only the correlation between 'involvement' and the overall PASS scores cannot be justified in this research study. Thus, out of the four individual cultural traits, those of 'mission', 'consistency' and 'adaptability' have a significant influence on organisational performance and 'involvement' has no significant influence. These results show that the traits of 'adaptability', 'consistency' and 'mission' are positively correlated and significantly or highly significantly associated with higher PASS scores in 'architectural and structural works' and to a lesser degree with other PASS scores such as 'other obligations', 'site management' and 'programme and progress'. The results also show that while the trait of 'involvement' correlates positively with 'architectural and structural works', it has no significant relationship to the other aspects of PASS performance. There does appear to be a positive association between the traits of 'consistency' and 'mission', and a high-ranking position in the PLS (where companies are actually in the PLS). Conclusions resulting from this are that 'adaptability', according to Denison and Neale (1994), is the organisational trait that demonstrates the capability of a company to receive, interpret and translate signals from its environment into internal behavioural changes that increase its chances for survival, growth and development, i.e. to turn the demands of the business environment into productive and effective action. Three aspects of adaptability impact upon an organisation's effectiveness:

- creating change;
- customer focus;
- organisational learning.

The high significance of the relationship between 'adaptability' and performance strongly underpins the fact that in a sector such as the construction industry, where the business environment is so volatile and subject to massive and often swift changes, the companies most able to survive are those that are most capable of adapting to these changes. This is particularly pertinent in relation to the public housing sector where housing production levels, cost yardsticks and competitiveness are so dependent on public policy, as well as being directed and driven by altered public sentiment, people's increasing aspirations, and media and political pressure. The significant relationship between the traits of 'consistency' and 'mission' and performance highlights the necessity for successful companies to possess core values that are understood by all players and well integrated throughout the organisation through commonly shared goals and objectives and a strong agreed strategic direction. These traits are strongly related to the stability of companies, which creates well-being, loyalty and satisfaction among staff. This stability must importantly be reflected in a long-term mission focus which although it may have to adapt, should not significantly change over short spans of time. Denison and Neale (1994: 356–357) state that 'The most troubled organisations are often those that have had to change their basic mission … when an organisation's underlying mission changes, corresponding changes in strategy, structure, culture, and behavior are also required.' The low significance of the relationship between 'involvement' and performance is an interesting and somewhat surprising result since, according to Denison (2000), organisational cultures characterised as 'highly involved' strongly encourage employee involvement and create a sense of ownership and responsibility. Their internal management systems are based on informal, voluntary and implied control rather than on formal, explicit or bureaucratic control, and strong commitment to an organisation develops from a sense of personal ownership and an increasing capacity for autonomy. Highly involved organisations seek input from their members in order to improve the quality of the decisions made and their subsequent implementation.

The indices of the 'involvement' trait (Denison *et al.*, 2004) on Denison's current website (2009) are:

- *Empowerment* – Individuals have the authority, initiative and ability to manage their own work. This creates a sense of ownership and responsibility towards the organisation.
- *Team orientation* – Value is placed on working cooperatively towards common goals for which all employees feel mutually accountable. The organisation relies on team effort to get work done.
- *Capability development* – The organisation invests continually in the development of employees' skills in order to stay competitive and meet ongoing business needs.

Reverting back to the four mini-case studies in Chapter 9, the 'best' performing companies certainly exhibited strong involvement characteristics and this accords with the results of Denison (1996, 2000) and others (Fey and Denison, 2000; Fisher, 2000; Bryant and Fleenor, 2002). However, the overall results may have been distorted by the fact that while many of the other companies surveyed aspire towards greater involvement, which would be demonstrated by high 'input' scores under PASS, their quality of work may not have been as adequate and thus their overall PASS scores would suffer greatly (PASS 'output' scores represent 70 per cent of overall composite scores). This would significantly affect the correlation results for this OC trait.

With reference to Research Question 3

Are any combinations of the four traits, based on a horizontal or vertical split of the Denison Organisational Culture Model, more significant in contributing to success levels than others?

From the results described in Chapter 8, it may be seen that all four combinations of the cultural traits correlate positively with the overall PASS score at the 0.05 significance level. However, comparing the results more closely, 'external focus' and 'stable culture' correlate more significantly than 'internal focus' and 'flexible cultures'. The purpose of examining these combined cultural trait combinations and their relationship with performance levels was to ascertain whether companies that could manage these apparently contradictory competing dimensions of the Denison Organisational Culture Model performed better than those companies that did not handle them well. Denison and Neale (1994: 357) states that 'Effective organisations find a way to resolve these dynamic contradictions without relying on a simple trade-off'. The trade-offs are:

- stability versus flexibility;
- internal versus external focus;

- internal consistency versus external adaptation;
- top-down mission versus bottom-up involvement.

Interestingly, the results also show that in companies possessing strong DOCS combined trait 'mission/consistency' (i.e. 'stable' company structure) and 'involvement/consistency' (i.e. 'internally focused' company) scores, there is a significant association with a high-ranking position in the PLS (where companies are in the PLS). There appears to be no significant relationship between DOCS combined trait 'involvement/adaptability' (i.e. flexible company structure) and 'adaptability/mission' (i.e. 'externally focused' company) scores and a high level of overall organisational performance measured by a high-ranking position in the PLS (where companies are in the PLS). As explained in Chapter 2, the PLS was developed to establish a specialised list of 'super'-contractors within HKHA's normal contractor lists. The PLS was intended to be formed of contractors with strong corporate direction, goal-motivated and capable of partnering with HKHA on complex or specialised high-value projects. As the attributes of 'mission' and 'consistency' locate within the 'stable' half of the Denison Organisational Culture Model, it is not surprising that companies with strong culture traits in this area are seen as being the best 'long-term' partners by HKHA. Stability indicates a mature organisation that knows where it is going and in which the major core values, goals and objectives and strategic direction are understood by all.

Implications for theory

The results of the research described in this book adds to the theory in the following areas:

- They support the outcomes of the work of other researchers who have investigated the presumed link between organisational culture and organisational performance, in that there appears to be a strong association between the two constructs.
- They use an entirely new set of measures to operationalise business success and performance (i.e. the HKHD's PASS scoring system) in a new sector of research (i.e. the construction industry) in which robust research effort is only recently beginning (CIB TG23/W112 initiatives).
- The DOCS instrument of Denison and Neale (1994) has undergone a rigorous process of decentring and back-translation to better adapt it for use in a new research environment: Hong Kong, construction, public sector housing.
- Support for mixed-method research approaches is provided by way of the use of simple qualitative enquiry to underpin quantitatively obtained data.

- These results posit the possibility to further develop the DOCS for specific use in the construction industry setting and as a longitudinal metric for change measurement in this dynamic industry.

Implications for the Hong Kong Housing Department and for companies engaged in public housing construction in Hong Kong

The Hong Kong Housing Department (HKHD)'s Performance Assessment Scoring System (PASS)

Certain findings of the research described in this book impact on the future development, improvement and use of PASS as an objective performance measurement tool. While this study has utilised PASS as its measure of organisational effectiveness, clearly certain aspects and components of the system appear not to demonstrate true organisational performance as well as others. While the aspects of 'architectural and structural works' and 'programme and progress' are indeed strong measures of effectiveness, being process based and highly objective, 'management input', 'safety' and 'other obligations' aspects of the system do not appear to represent performance so accurately, since they are based on more subjective measurement methodologies. Using the now-proven significant relationship between the culture strength measures under DOCS and the performance scores obtained under PASS, the latter could be reviewed and redeveloped so that the strong objective measures (i.e. 'architectural and structural works' and 'programme and progress') could be combined with equally strong empirical measures of what the HKHD may currently regard as 'softer' aspects of a company's organisational structure such as its culture.

Construction companies

The interest of the Hong Kong construction industry through ongoing discussions with the Hong Kong Contractors' Association (HKCA), and of individual contractors, in the results of this study has been maintained throughout the time of conducting the research and during the writing of the subsequent thesis that followed, and which was completed in 2006. Many companies have since expressed an interest in knowing more about their organisational culture profile and particularly how they benchmark against the industry norm and the perceived 'best' performers in the industry. Some companies that have made significant organisational changes recently have requested that the author considers providing a further survey exercise for them to compare their results before and after change. This latter request is a worthy foundation for future research and will currently be discussed further with these companies to establish how further longitudinal research could be operated and when would be the best time to carry out the study. In order to satisfy the former requests and in accordance with promises made to HKCA and the participating respondent companies, the author developed a short executive summary of the research results based on a paper presented at the Curtin University Graduate Business School Doctoral Colloquium in 2003, which sets out the general results of the original thesis and advocates those traits of organisational culture that appear most closely related to high performance in respect of public sector housing projects. This was then made available to individual contractors upon request.

In an industry and location (i.e. construction and Hong Kong) which has not until recently had a successful history of being able to benchmark itself on a company, sector or global basis, the tremendous interest which has been generated by the original research study indicates that a new set of tools for the industry to gauge its efficacy and progress during and after change is long overdue. DOCS and PASS are a good combination to perhaps drive the further development and production of these much sought-after tools.

Implications for further research

The organisational culture (OC) and organisational performance (OC) link

While much has been achieved by way of research into the OC–OP link since the late 1980s, many authors share the view expressed by Sparrow (2001: 102) that,

> in taking the more functionalist and positivist perspective of high performance culture research, it is important to stress that this form of assessment [i.e. the OC–OP research perspective] need not be limited to quantitative methodologies. Qualitative tools and techniques can sit just as easily in the consultant's toolkit ... there must be room for culture and climate diagnostics, informed by richer – and in some instances more qualitative – investigation, that tap the most relevant individual sense making processes.

Certainly, the qualitative investigation carried out for the research that has been described in detail in this book clearly indicated a rich source of alternative data to better inform the purely quantitative results of using an organisational culture measuring instrument, and this accords with the view of the said instrument's author, who stated (Denison, 2001: 367): 'Perceptive insiders and outsiders need to be involved in order to help translate the findings from a model-based analysis of the culture...depth of analysis is needed to support the insights from the survey data and bring them to life.' Clearly, then, there is a need for more qualitative work to be linked with quantitative studies to increase our knowledge of organisational structure, culture and its affect on performance, and although current studies are beginning to move us more swiftly in this direction, further work is required to strengthen this whole field of study.

The metrics of organisational performance

Hopefully, this book has clearly demonstrated the benefits of considering alternative ways to operationalise organisational performance other than by the use of purely financial measures. As mentioned in Chapter 4, Kennerley and Neely (2002) summarised the main components of an effective performance measurement system as follows:

- must provide a 'balanced' picture of the business;
- needs to present a succinct overview of the organisation's performance;
- should be multidimensional;
- requires comprehensiveness;
- must be integrated both across the organisation's functions and through its hierarchy;
- business results need to be seen as a function of the measured determinates.

Kennerley and Neely's own integrated and stakeholder-focused performance measurement model, the Performance Prism (2002), appears to satisfy the parameters described above and should be more widely used, together with other instruments such as the Balanced Score-card (Kaplan and Norton, 1992), in order to become more universally accepted measures of organisational effectiveness. This would then allow the development of local, national and global performance databases for use by both researchers in academic studies and by businesses themselves for benchmarking and setting up strategic organisational change programmes.

The Hong Kong construction industry

Researching the culture of the 23 companies involved in constructing public sector housing in Hong Kong has been a first step only into an area in which a lot more research is clearly needed (Root, 1994; Chinowsky and Meredith, 2000; Fellows and Seymour, 2002), and certainly the results described in this book need to be extended both laterally (i.e. launched into construction companies outside public sector housing) and longitudinally (i.e. how the organisational cultures of companies surveyed using DOCS have changed since the time of the original study and what affect these changes have had on organisational performance as a result). Through the efforts of the CIB Task Group TG23 (now Working Group W112), research into organisational and corporate culture in the construction industry is now progressing, and this will be at a global as well as a national level of consideration. It is interesting to note in this context that the chosen instrument of culture measurement is the Organisational Culture Inventory (Cooke and Rousseau, 1983) described in Chapter 3. This methodology promises to build up a global database of cultural profiles of construction organisations that is user-friendly and highly usable at a national and global level for research purposes.

The last word

The original thesis from which this book was developed was conceived from the researcher's interest and belief in the existence of a demonstrable link between the cultures of organisations and the impact that different levels of cultural strength may have upon the effectiveness and performance of these organisations. Building on the theory which has developed out of the many studies carried out in the past decade or so that have demonstrated such a link exists (Denison, 1990; Gordon and DiTomaso, 1992; Kotter and Heskett, 1992; Chapman and Connell, 1995; Wilderom and van den Berg, 1998), while at the same time recognising the important contribution of other studies which have been critical of the validity of such presumptions (Saffold, 1988; Siehl and Martin, 1990; Lim, 1995; Wilderom *et al.*, 2000), this research thesis was designed to test both sets of views.

A large part of the doubt expressed by the latter group of authors related to the measures of organisational culture and effectiveness currently being used in such research, in particular the reliance on financially based measures of effectiveness rather than other tangible metrics. This discussion on differing theoretical opinions, methodologies and measurement tools for investigating the existence of the culture–performance link has presented a fascinating opportunity for further research and, given the author's long-time experience and specific placement in the construction industry since 1977, a topic developed that not only examined the organisational culture and performance link, but carried the research specifically into construction industry organisations and public sector building projects, which, according to some authors (e.g. Chinowsky and Meredith, 2000), is a sector of business where a need exists for more research into those management and organisational issues that impact on the efficacy of construction companies to perform more effectively. The basic research questions addressed throughout the chapters of this book are: Why are some companies able to produce built output that better satisfies their customers when many others cannot; and Does the organisational culture of the construction company impact on its performance? Using the list of new works building contractors of the Hong Kong Housing Authority (HKHA), secondary data drawn from that organisation's objective quality measurement tool known as the Performance Assessment Scoring System (PASS) as an alternative objective measure of effectiveness and the Denison Organisational Culture Survey (DOCS) instrument developed in 1984 by Dr Daniel Denison, research was then undertaken on the relationship between organisational culture and organisational performance. The outcome of the research was that a link does indeed exist between the two and that there is also significant correlation between the strength of an organisation's culture and its comparative effectiveness in performance terms when investigated in the specific setting of Hong Kong. This conclusion, when it was developed in the thesis that motivated the writing of this book, does make a contribution to theory by further validating the work by Denison (1990) and others, not only in that it successfully demonstrates a link between organisational culture and performance, but it also contributes to management and public policy literature by identifying particular cultural factors in organisations that appear to be significantly responsible for achieving successful outcomes, and reveals opportunities for further research into the organisational culture of construction companies.

Appendix A

Summary of the organisational culture instruments in the derivation of the OCP

Author	Year	Level	Dimensions	Type	Format	Reliability	Consensual validity	Construct validity	Criterion validity
Harrison	1975	½	15	Typing	Rank				
Handy	1979	2	9	Typing	Rank				
Margerison	1979	2	24	Typing	Likert				
Organisation Technology International	1979	2	5	Descriptive	Likert				
Allen & Dyer	1980	1	7	Behaviour[b]	Likert			X	X
Kilmann & Saxton	1983	1	4	Behaviour[b]	Paired	X		X	X
Glaser	1983	2	4	Typing	Likert				
Harris & Moran	1984	2	7	Effectiveness	Likert				
Sashkin & Fulmer	1985	2	10	Effectiveness	Likert		X		
Cooke & Lafferty	1986	1	12	Behaviour[b]	Likert	X	X	X	X
Enz	1986	2	22	Fit	Likert	X		X	X
Reynolds	1986	2	14	Descriptive	Likert	X			X
Woodcock	1989	2	12	Effectiveness	Likert				
Hofstede et al.	1990	2	N/A	Descriptive	Likert				
Lessem	1990	2	7	Typing	Rank				
O'Reilly et al.	1991	2	24	Fit	Q-sort	X	X	X	X
PA Consulting Group	1991	2	N/A	Descriptive	Likert				
Migliore et al.	1992	2	20	Descriptive	Likert		X		

Ashkanasy et al. (2000) 'Level column': 1 = Patterns of behaviour, 2 = Values and beliefs (Schein, 1985). Behaviour[b] = Patterns of behaviour (Note: the table has been sorted chronologically, a deviation from the original).

Appendix B

Organizational Culture Survey (back-translated version for use in Hong Kong)

Management practices and organizational strategies are rooted in the underlying beliefs, values, and assumptions held by members of an organization. The approach that underlies the Denison Organizational Culture Survey is based on a model of four cultural traits of organizations. These traits have been linked by research to specific aspects of performance and effectiveness such as return on assets, quality, sales growth, and employee satisfaction.

This survey presents a set of 60 statements that describe different aspects of an organization's culture and ways that organizations operate. To complete the survey, just indicate how much you agree or disagree with each of the statements. When you are responding to the statements, think of your organization as a whole and the way that things are usually done. If the statement is a good description of the way that things are typically done in your organization, then you should indicate that you agree with that statement. If the statement is not a good description of the way things typically work in your organization, then indicate that you disagree.

Using the five descriptive response categories indicated below, please select the appropriate choice next to each statement in the answer column by placing a *tick* in the box to the left of the number corresponding to the choice that you agree with the most.

Instructions

Please respond to every item listed by selecting the number that best describes the culture of your organization. If you have difficulty deciding between two response options, select the one that you think most closely describes the organization. Neutral should be used when you are stuck in the middle and neither agree nor disagree with the statement. In cases where an item is not applicable, leave the individual item blank.

ID	OPINIONS
	In this organization...
1	Most employees are committed to their jobs.
2	When people are provided with adequate information, they usually can make decision easily.
3	Everyone can get his information as soon as he needs provided that information is shared among him or her.
4	Everyone believes that he or she can make positive impact to the company.
5	Commercial planning is continuously ongoing and everyone will take part in this planning to some extent.
6	Cooperation between different departments is highly recommended by company.
7	People work together as teams.
8	Teamwork, rather than hierarchy, helps to get tasks completed.
9	Teams are essential elements of our work.
10	Jobs are allocated so that people can realize the relationship between their job and those of others and have a better understanding of the company's goals.
11	People are authorized to do their jobs without having to seek further authority from above.
12	The capability of people to do their job well is continuously improving.
13	The company will continuously invest in its staff's techniques.
14	The company's competitive advantage over others depends on the capability of its staff.
15	Problems arise because we lack the techniques required to carry out our work properly.
16	Managers and leaders should have consistency between what they speak and practice.
17	The company has a unique management style and distinct management practices.
18	The company has a clear and consistent set of values to govern the way we conduct business.
19	Troubles arise when workers ignore the core values.
20	The company has a set of ethical codes to guide its staff's behaviour and help them to identify right from wrong.
21	When controversy arises, we will work our best to achieve an agreement that satisfies all sides.
22	The company has its own main culture.
23	We can easily reach consensus on different issues, no matter how difficult they are.
24	It is usually more difficult to reach consensus on key issues.
25	There is a clear agreement to decide the right and wrong way to work.
26	Our approach to doing business is very consistent and can easily be anticipated.

(Continued)

ID	OPINIONS
27	Staffs of different departments have the same points of view.
28	It is easy to coordinate projects among different departments.
29	When we work with colleagues from other departments in the same company, we feel like we are working with people in another company.
30	There are common targets across levels.
31	The ways we do business are flexible and easy to change.
32	Faced with changes by competitors to their business environment, the company can have good responses.
33	The company can continuously adopt the latest and improved ways to work.
34	We usually encounter obstacles when we attempt to change things.
35	Different departments will help each other to create changes.
36	The customer's opinion and recommendations are usually the keys to our changes.
37	The customer's involvement will directly influence the company's decision.
38	Everyone in the company has a deep understanding of the customers' needs.
39	Company will usually ignore the interests of the customers when decisions are made.
40	We encourage direct interaction between staff and customers.
41	The company view's failure as a chance to learn and improve.
42	The company encourages innovations and risk-taking and gives rewards if these are seen.
43	Many things which are important go unnoticed.
44	Learning is one of the main goals in everyday work.
45	We make sure that there are understandings between the boss and the staff.
46	The company has long-term goal and directions.
47	The company's strategy can guide other organizations to change the way they compete in this industry.
48	The company's clear mission brings meaning and direction to our works.
49	The company has a clear strategy for the future.
50	I am not clear as to the company's strategic direction.
51	There is widespread agreement about the company's goal.
52	The goal being set by the leaders is ambitious and achievable.
53	Leaders have formally communicated records of the business goal.
54	We will continuously track our progress to the goal we set.
55	People will know how to achieve long-term goals.
56	We have the same vision of what the company will be like in the future.
57	Leaders have a long-term viewpoint.
58	Our short-term thinking often undermines the long-term vision.
59	Our vision creates excitement and motivation for our employees.
60	We can still face short-term needs without having compromise with the long-term viewpoint.

Denison Organizational Culture Survey - designed by Daniel R. Denison Ph.D. and William S. Neale MA.MLIR.

Note: The survey was developed in USA and has been adapted by the researcher to better suit the Hong Kong environment.

The following items will be used to evaluate trends within your organization and contribute greatly to the accuracy of the research. Your responses will be treated **confidentially** and will *never* be used to identify specific individuals.

Age		Organizational level	
☐	Under 20	☐	Non-exempt/non-management
☐	20–29	☐	Exempt/line management (supervising non-management)
☐	30–39	☐	Middle management (managing managers)
☐	40–49	☐	Senior management
☐	50–59	☐	Executive/senior vice-president
☐	60 or over	☐	CEO/president

Sex
- ☐ Female
- ☐ Male
- ☐ Prefer not to respond

Education (mark highest level)
- ☐ Secondary school
- ☐ Some college education
- ☐ Technical diploma
- ☐ Bachelor's degree
- ☐ Some graduate work
- ☐ Master's degree
- ☐ Doctoral degree
- ☐ Other

Department
- ☐ Finance and Accounting
- ☐ Engineering
- ☐ Manufacturing and Production
- ☐ Research and Development
- ☐ Sales and Marketing
- ☐ Purchasing
- ☐ Human Resources
- ☐ Administration
- ☐ Support staff
- ☐ Professional staff

Years With organization
- ☐ Less than 6 months
- ☐ 6 months to 1 year
- ☐ 1 to 2 years
- ☐ 2 to 4 years
- ☐ 4 to 6 years
- ☐ 6 to 10 years
- ☐ 10 to 15 years
- ☐ More than 15 years

Salary (Annual) in HK Dollars
- ☐ $25,000 or less
- ☐ $25,001 to $35,000
- ☐ $35,001 to $50,000
- ☐ $50,001 to $75,000
- ☐ $75,001 to $100,000
- ☐ $100,001 to $150,000
- ☐ $150,001 to $200,000
- ☐ $200,001 plus

Ethnic Background
- ☐ Asian
- ☐ African American
- ☐ Hispanic
- ☐ White/Caucasian
- ☐ Other

The following set of questions asks about the performance of your organization. Compared to companies like yours, how would you assess your organization's performance in the following areas?

	Low performer	Average	High performer
Sale/revenue growth	☐	☐	☐
Market share	☐	☐	☐
Profitability/ROA	☐	☐	☐
Quality of products or services	☐	☐	☐
New product development	☐	☐	☐
Employee satisfaction	☐	☐	☐
Overall organization performance	☐	☐	☐

The following set of questions asks about your perceptions of the significance of the concepts of 'value' within your organization.

	Question	Answer
1	What makes me feel most valued?	[Enter text here]
2	What is the thing I most value?	[Enter text here]

Thank you for completing the survey and contributing significantly to an important piece of research. As mentioned in the covering letter, which accompanies this survey, in due course a copy of your company's 'Organizational Culture Profile' will be sent to you and this will be benchmarked to a public housing sector construction industry set of 'norms', which will provide you with valuable feedback on how your company compares in various aspects of organizational culture with others in your sector. Please take a moment to read the note below regarding consents, anonymity and confidentiality.

IMPORTANT NOTE: Completion and return of this survey will be taken as representing your formal *CONSENT* to allow the data so gained to be included *ANONYMOUSLY* in the research thesis; and also to allowing, 'in principle', follow-up interviews to be undertaken at a later stage of the research. All information obtained during the survey and any subsequent interviews will be treated *CONFIDENTIALLY* and no company will be specifically or individually identified in the research thesis text.

References

Abegglen, J. C. & Stalk, G., Jr. 1986, *Kaisha: The Japanese Corporation*. Basic Books, New York.

Abeysekera, V. 2003, "Understanding culture in an international construction context", in *CIB Publication 275*, R. F. Fellows & D. E. Seymour, eds. International Council for Research and Innovation in Building and Construction, Rotterdam, Netherlands, p. 51.

Accel-Team. 2003, *Elton Mayo's Hawthorne Experiments*. Available online HTTP: <http://www.accel-team.com/human_relations/hrels_01_mayo.html> (accessed 5 July 2009).

Aceves, J. & King, H. 1978, *Cultural Anthropology*. General Learning Press, Morristown, NJ.

Ahmed, S. M., Tang, P., Azhar, S. & Ahmad, I. 2002, "An evaluation of safety measures in the Hong Kong construction industry based on total quality management techniques", CRC Press, LLC. Boca Raton, FL; University of Cincinatti, Ohio.

Akin, G. & Hopelain, D. 1986, "Finding the culture of productivity", *Organizational Dynamics*, vol. 7, no. 2, pp. 19–32.

Aldrich, H. E. 1979, *Organizations and Environments*. Prentice-Hall, Englewood Cliffs, NJ.

Allen, J. & Allen, R. F. 1990, "A sense of community, a shared vision, and a positive outlook: key enabling factors in successful culture change", in *The Corporate Culture Sourcebook*, R. Bellingham et al., eds. Human Resource Development Press, Amherst, MA.

Allen, M. P., Panian, S. K. & Lotz, R. E. 1979, "Managerial succession and organizational performance: a recalcitrant problem revisited", *Administrative Science Quarterly*, vol. 24, pp. 65–81.

Alvesson, M. 1993, *Cultural Perspectives on Organizations*. Cambridge University Press, Cambridge, UK.

Alvesson, M. 2002, *Understanding Organizational Culture*. Sage, London.

Alvesson, M. & Berg, P. 1992, *Corporate Culture and Organizational Symbolism*. Walter de Gruyter., Berlin.

Anderson, N. & West, M. A. 1998, "Measuring climate for work group innovation: development and validation of the Team Climate Inventory", *Journal of Organizational Behaviour*, vol. 19, pp. 235–258.

Argyris, C. 1960, *Understanding Organizational Behavior*. Dorsey, Homewood, IL.

Argyris, C. 1964, *Integrating the Individual and the Organization*. John Wiley & Sons, New York.

Arnold, J., Cooper, C. and Robertson, I. 1995, *Work Psychology*, 2nd edn. Pitman, London, UK.

Ashforth, B. 1985, "Climate formation: issues and extensions", *Academy of Management Review*, vol. 10, pp. 837–847.

Ashkanasy, N. M., Broadfoot, L. & Falkus, S. 2000a, "Questionnaire measures of organizational culture", in *Handbook of Organizational Culture and Climate*, N. M. Ashkanasy, C. P. M. Wilderom & M. F. Peterson, eds. Sage, Thousand Oaks, CA, pp. 131–146.

Ashkanasy, N. M., Wilderom, C. & Peterson, M. F. 2000b, *Handbook of Organizational Culture & Climate*. Sage, Thousand Oaks, CA.

Badri, M. A., Davis, D. & Davis, D. 1995, "A study of measuring the critical factors of quality management", *International Journal of Quality & Reliability Management*, vol. 12, no. 2, pp. 36–53.

Baltes, B. B. & Parker, C. P. 2000, "Reducing the effects of performance expectations on behavioral ratings", *Organizational Behavior and Human Decision Processes*, vol. 82, pp. 237–267.

Baltes, P. 1997, 'On the incomplete architecture of human ontogeny: selection, optimization, and compensation as foundations of developmental theory', *American Psychologist*, vol. 52, no. 4, pp. 366–380.

Barforth, S., Duncan, R. & Miller, C. 1999, "A literature review on studies in culture: a pluralistic concept", in *Profitable Partnering in Construction Procurement*, S. Ogunlana, ed. E and FM Spon, London, UK, pp. 533–542.

Barker, R. 1994, "Relative utility of culture and climate analysis to an organizational change agent: an anlysis of general dynamics, electronics division", *International Journal of Organizational Analysis*, vol. 2, no. 1, pp. 68–87.

Barley, S. R., Meyer, G. W. & Gash, D. C. 1988, "Cultures of culture: academics, practitioners and the pragmatics of normative control", *Administrative Science Quarterly*, vol. 33, pp. 24–59.

Barnard, C. I. 1938, *The Functions of an Executive*, 1st edn. Harvard University Press, Cambridge, MA.

Barney, J. 1986, "Organizational culture: can it be a source of sustained competitive advantage?", *Academy of Management Review*, vol. 11, pp. 656–665.

Bates, D. G. & Plog, F. 1990, *Cultural Anthropology*, 3rd. edn. McGraw-Hill Education, New York.

Battersby, J. R. P. 2004, *Management of Construction Risk. Summary of a Talk Given to the Singapore Contractors Association, Ltd.* Available online HTTP: <http://www.scal.com.sg/index.cfm?GPID=106 > (accessed 5 July 2009).

Bavendam Research Incorporated. 2003, *Special Reports no. 6 – Managing Job Satisfaction.* Available online HTTP: <http://www.dtic.mil/cgi-bin/GetTRDoc?AD=ADA421924&Location=U2&doc=GetTRDoc.pdf> (accessed 5 July 2009).

Bedeian, A. G. 1987, "Organization theory: current controversies, issues and directions", in *International Review of Industrial and Organizational Psychology*, C. L. Cooper & I. T. Robertson, eds. John Wiley & Sons, New York, pp. 1–33.

Benedict, R. 1934, *Patterns of Culture*. Houghton Mifflin., Boston, MA.

Benjamin, L., Goodyear, L, & Greene, J. C. 2001, "The merits of mixing methods in evaluation", *Evaluation*, vol. 7, no. 1, pp. 25–45.

Bennett, J., Flanagan, R. & Norman, G. 1987, *Capital & Counties Report: Japanese Construction Industry*. Centre for Strategic Studies in Construction, University of Reading, Working Paper 21.

Berelson, B. 1952, *Content Analysis in Communication Research*. Free Press., Glencoe, ILL.

Berg, P. 1985, "Organizational change as a symbolic transformation process," in *Organizational Culture*, P. J. Frost et al., eds. Sage, Beverly Hills, CA, pp. 281–299.

Berry, J. W. 1969, "On cross-cultural comparability", *International Journal of Psychology*, vol. 4, pp. 119–128.

Berry, J. W. 1990, "Imposed etics, emics, derived etics: their conceptual and operational status in cross-cultural psychology", in *Emics and Etics: The Insider/Outsider Debate*, T. N. Headland, K. L. Pike & M. Harris, eds. Sage, Newbury Park, CA, pp. 28–47.

Bhargava, S. & Sinha, B. 1992, "Predictions of organizational effectiveness as a function of type of organizational structure", *Journal of Social Psychology*, vol. 132, pp. 223–232.

Binford, L. R. 1968, "Archaeological perspectives", in *New Perspectives in Archaeology*, Aldine, Chicago, IL.

Bluedorn, A. C. 1980, "Cutting the Gordian knot: a critique of the effectiveness tradition in organizational research", *Sociology and Social Research*, vol. 64, pp. 477–496.

Boaz, F. 1940, *Race, Language and Culture*. Macmillan, New York.

Bodley, J. H. 2000, *Cultural Anthropology: Tribes, States, and the Global Systems*, 3rd edn. Mayfield, Mountain View, CA.

Boer, P. *Work Values Checklist*. Monster Worlwide, Inc. Available online HTTP: <http://content.comcast.monster.com/selfassessment/Work-Values-Checklist/home.aspx> (accessed 5 July 2009).

Borofsky, R. 1993, *Assessing Cultural Anthropology*. McGraw-Hill Higher Education, New York.

Bourgeois, L. J. I. 1980, "Performance and consensus", *Strategic Management Journal*, vol. 1, no. 227, p. 248.

Bourgeois, L. J. & Eisenhardt, K. 1988, "Strategic decision processes in high velocity environments: four cases in the micro-computer industry", *Management Science*, vol. 34, pp. 816–835.

Bourne, M. 2001, *The Handbook of Performance Measurement*. Gee Publishing, London, UK.

Bowman, J. 1980, "The importance of a market value measurement of debt in assessing leverage", *Journal of Accounting Research*, vol. 18, pp. 242–254.

BRE. 2000, *The Housing Quality Indicators*. Building Research Establishment, Garstang, UK.

Bresnen, M. & Marshall, N. 2000, "Partnering in construction: a critical review of issues, problems and dilemmas", *Construction Management and Economics*, vol. 18, no. 2, pp. 229–237.

Bresnen, M. & Marshall, N. 2000, "Motivation, commitment and the use of incentives in Partnerships and Alliances.", *Construction Management and Economics*, vol. 18, no. 5, pp. 587–598.

Brett, J. M., Tinsley, C. H., Janssens, M., Barsness, Z. I. & Lytle, A. L. 1997, "New approaches to the study of culture in industrial/organizational psychology", in *New Perspectives on International Industrial/Organizational Psychology*, P. C. Earley & M. Erez, eds. New Lexington Press, San Francisco, CA, pp. 75–129.

Brewer, J. & Hunter, A. 1989, *Multimethod Research: A Synthesis of Styles*. Sage, Newbury Park, CA.

Breyfogle, F., Cupello, J. & Meadows, B. 2001, *Managing Six Sigma: A Practical Guide to Understanding, Assessing, and Implementing the Strategy that Yields Bottom-line success*. John Wiley, New York.

Brignall, T. J. S. 2003, The unbalanced scorecard: a social and environmental critique. Boston, MA, unpublished paper under review at *International Journal of Operations and Production Management*.

Brislin, R. W. 1976, "Comparative research methodology: cross-cultural studies", *International Journal of Psychology*, vol. 11, no. 3, pp. 215–229.

Brislin, R. W. 1993, *Understanding Culture's Influence on Behaviour*, 1st edn. Harcourt Brace College Publishers, Fort Worth, TX.

Brislin, R. W., Lonner, W. & Thorndike, R. 1973, *Cross-cultural Research Methods*. John Wiley, New York.

British Standards Institute. 1992, *Guide to Management Principles, BS 7850 Part 1*. British Standards Institute, London, UK.

Broadfoot, L. and Ashkanasy, N. M. 1994, 'A survey of organisational culture measurement instruments'. Paper presented at the 23rd Meeting of Australian Social Psychologists, Cairns, Australia.

Bröchner, J., Josephson, P. E. & Kadefors, S. 2002, "Swedish construction culture, quality management and collaborative practice", *Building Research and Information*, vol. 30, no. 6, pp. 392–400.

Brooks, J. G. & Brooks, M. G. 1993, *In Search of Understanding: The Case of the Constructivist Classroom*. ACSD, Alexandria, VA.

Brooks, R. to Landrum, J. (email). 2002, Boasian tradition in Bates and Plog definition of culture.

Brown, A. D. 1998, *Organisational Culture*, 2nd edn. Financial Times, London.

Brown, M. & Laverick, S. 1994, "Measuring corporate performance", *Long Range Planning*, vol. 27, no. 4, pp. 89–98.

Brown, M. G. 1996, *Keeping Score: Using the Right Metrics to Drive World-class Performance*. Quality Resources, New York.

Brown, W. B. 1998, *The Influence of Organization Culture on Organization Performance*. Dissertation Abstracts International, 59 (02), 612A (UMI no. 9825762).

Bruns, W. 1998, "Profit as a performance measure: powerful concept, insufficient measure", Performance Measurement - Theory and Practice; The First International Conference on Performance Measurement, Cambridge, pp. 14–17.

Bryant, C. and Fleenor, W.J. 2002, 'Leadership effectiveness and organizational cultural: An exploratory study'. Unpublished paper presented at the meeting of Industrial and Organizational Psychology, Toronto, Canada, April.

Buckley, W. F. 1968, "Society as a complex adaptive system", in *Modern Systems Research for the Behavioral Scientist*, W. F. Buckley, ed. Aldine, Chicago, IL, pp. 490–513.

Burrell, G. & Morgan, G. 1992, *Sociological Paradigms and Organizatioanl Analysis*. Heinemann, Portsmouth, NH.

Business Week. 1980, Corporate culture: the hard-to-change values that spell success or failure, pp. 148–160. 27 October.

Business Week. 1984, Who's excellent now?, pp. 46–55. 11 May.

Byrne, J. A. 2001, *The Real Confessions of Tom Peters*. Business Week 3 December. Available online HTTP: <http://www.fastcompany.com/magazine/53/peters.html> (accessed 5 July 2009).

Calás, M. & Smircich, L. Post-culture: is the organizational culture literature dominant but dead? Paper presented at the International Conference on Organizational Symbolism and Corporate Culture, Milan.

Calori, R. & Sarnin, P. 1991, "Corporate culture and economic performance: a French study", *Organization Studies*, vol. 12, pp. 49–74.

Cameron, K. S. 1980, "Critical questions in assessing organizational effectiveness", *Organizational Dynamics*, vol. 8, autumn, pp. 66–80.

Cameron, K. S. 1981, "Domains of organizational effectiveness in colleges and universities", *Academy of Management Journal*, vol. 24, pp. 25–47.

Cameron, K. S. 1984, "The effectiveness of ineffectiveness", *Research in Organizational Behavior*, vol. 6, pp. 235–285.

Cameron, K. S. 1986, "Effectiveness as a paradox: consensus and conflict in conceptions of organizational effectiveness", *Management Science*, vol. 32, pp. 539–553.

Cameron, K. S. & Freeman, S. 1989, "Cultural congruence, strength, and type: relationships to effectiveness'. Presentation to the Academy of Management Annual Convention. Academy of Management, Washington, DC.

Cameron, K. S. & Quinn, R. E. 1999, *Diagnosing and Changing Organizational Culture Based on the Competing Values Framework*. Addison-Wesley, Reading, MA.

Cameron, K. S. & Whetten, D. A. 1983, *Organizational Effectiveness: A Comparison of Multiple Models*. Academic Press, New York.

Campbell, D. T. & Werner, O. 1970, "Translating, working through interpreters and the problem of decentering", in *A Handbook of Method in Cultural Anthropology*, R. Naroll & R. Cohen, eds. Natural History Press, New York, pp. 398–420.

Carroll, D. T. 1983, "A disappointing search for excellence", *Harvard Business Review*, vol. 61, no. 6, pp. 78–88.

Cassidy, M. P. 1996, "Streamlining TQM", *The TQM Magazine*, vol. 8, no. 4, pp. 24–28.

CE - Construction Best Practice. 2003, *Construction - A Sector Study*. Constructing Excellence, Stafford, UK.

Chan, A. P. C. & Tam, C. M. 2000, "Factors affecting the quality of building projects in Hong Kong", *International Journal of Quality & Reliability Management*, vol. 17, nos 4/5, pp. 423–441.

Chapman. 2009, Available online HTTP: <http://www.businessballs.com/dtiresources/TQM_implementation_blueprint.pdf> (accessed 29 August 2009).

Cheng, E., Li, H. & Love, P. E. D. 2000, "Establishment of critical success factors for construction partnering", *Journal of Management in Engineering*, vol. 16, no. 2, pp. 84–92.

Cheung, P. C., Conger, A. J., Hau, K., Lew, W. J. F. & Lau, S. 1992, "Development of the multi-trait personality inventory (MTPI): comparison among four Chinese populations", *Journal of Personality Assessment*, vol. 59, no. 3, pp. 528–551.

Chin, K. S. & Pun, K. F. 2002, "A proposed framework for implementing TQM in Chinese organisations", *International Journal of Quality and Reliability Management*, vol. 19, no. 3, pp. 272–294.

Chinowsky, P. S. & Meredith, J. E. 2000, "Strategic management in construction", *Journal of Construction Engineering and Management*, vol. 17, no. 2, pp. 60–68.

Cho, H. J. 2000, The validity and reliability of the Organizational Culture Survey. Unpublished Working Paper.

Cicourel, A. V. 1974, *Cognitive Sociology*. Free Press, New York.

CIRC. 2001, *Construct for Excellence*, a report of the Construction Industry Review.

Clark, B. 1970, *The Distinctive College*. Aldine, Chicago, IL.

Clark, B. 1980, *Academic Culture*. Yale Higher Education Research Group, New Haven, CT.

Clegg, S., Hardy, C. & Nord, W. R. 1996, *Handbook of Organization Studies*. Sage, London.

Coffey, A., Holbrook, B. & Atkinson, P. 2009, *Qualitative Data Analysis: Technologies and Representations*. School of Social and Administrative Studies, University of Wales, Cardiff, UK. Available online HTTP: <http://www.socresonline.org.uk/1/1/4.html> (accessed 5 July 2009).

Coffey, W. V. 2002, "Assessing organizational culture? Value management to the rescue". Paper presented to the Institute of Value Management of Australia Conference, Tasmania, Australia.

Coffey, W. V. 2003, "The organisational culture and effectiveness of companies involved in public sector housing construction in Hong Kong - preliminary research results". Paper presented to the CIB TG23 International Conference "Professionalism in Construction: Culture of High Performance', Hong Kong.

Coffey, W. V. 2005, *The Organisational Culture and Effectiveness of Companies involved in Public Sector Housing Construction in Hong Kong*. DBA Thesis, Graduate School of Business, Curtin University, Perth, Australia.

Connolly, T., Conlon, E. J. & Deutsch, S. J. 1980, "Organizational effectiveness: a multiple-constituency approach", *Academy of Management Review*, vol. 5, pp. 211–217.

Constructing Excellence in the Built Environment. 2009, *Partnering and SCM - Construction Best Practice*. Available online HTTP: http://www.constructingexcellence.org.uk/resources/themes/external/management.jsp (accessed 5 July 2009).

Cooke, R. A. & Lafferty, J. C. 1983, *Organizational Culture Inventory (Form I)*. Human Synergistics, Plymouth, MI.

Cooke, R. A. & Lafferty, J. C. 1986, *Organizational Culture Inventory (Form III)*. Human Synergistics, Plymouth, MI.

Cooke, R. A. & Rousseau, D. M. 1983, "The factor structure of Level I: life styles inventory", *Educational and Psychological Measurement*, vol. 43, pp. 449–457.

Cooke, R. A. & Rousseau, D. M. 1988, "Behavioural norms and expectations: a quantitative approach to the assessment of organizational culture", *Group and Organization Studies*, vol. 13, no. 3, pp. 245–273.

Cooke, R. A. & Szumal, J. L. 1993, "Measuring normative beliefs and shared behavioural expectations in organizations: the reliability and validity of the Organizational Culture Inventory", *Psychological Reports*, vol. 72, no. 3, pp. 1299–1330.

Cooke, R. A. & Szumal, J. L. 2000, "Using the Organizational Culture Inventory to understand the operating cultures of organizations", in *Handbook of Organizational Culture and Climate*, N. M. Ashkanasy, C. P. M. Wilderom & M. F. Peterson, eds. Sage, Thousand Oaks, CA.

Cooley, C. H. 1922, *Human Nature and Social Order*. Scribner's, New York.

Coolican, H. 1990, *Research Methods and Statistics in Psychology*. Hodder & Stoughton, London, UK.

Cooper, D. R. & Schindler, P. S. 2000, *Business Research Methods*, 7th edn. McGraw-Hill Irwin, New York.

Coser, L. A. 1977, *Masters of Sociological Thought: Ideas in Historical and Social Context*. Harcourt Brace Jovanovich, New York.

Coulter, P. B. 1979, "Organizational effectiveness in the public sector: the example of municipal fire protection", *Administrative Science Quarterly*, vol. 24, pp. 167–180.

Cox, B. J., Ross, L., Swinson, R. P. & Direnfield, D. M. 1998, "A comparison of social phobia outcome measures in cognitive-behavioral group therapy", *Behavior Modification*, vol. 22, pp. 285–297.

Cronbach, L. J. 1951, "Coefficient alpha and the internal structure of tests", *Psychometrika*, vol. 16, pp. 297–344.

Cultural Research Inc. 2004, *Kent State University Workplace Culture Project - April 1998*. Available online HTTP: < http://dept.kent.edu/ksucultural_audit/> (accessed 5 July 2009).

Cunningham, J. B. 1977, "Approaches to the evaluation of organizational effectiveness", *Academy of Management Review*, vol. 2, pp. 463–474.

Cunningham, W. G. & Gresso, D. W. 1994, *Cultural Leadership: The Culture of Excellence in Education*. Allyn & Bacon, Boston, MA.

Czarniawska-Joerges, B. 1992, *Exploring Complex Organizations: A Cultural Perspective*. Sage, Newbury Park, CA.

D'Andrade, R. G. 1996, "Culture", in *The Social Science Encyclopedia*. Routledge, London, UK.

Dachler, H. P. 1997, "Qualitative methods in organizational research: a practical guide", *Organization Studies*, vol. 18, no. 4, pp. 709–724.

Daft, R. L. 2001, *Organization Theory and Design*, 7th edn. South-Western College Publishing, Mason, OH.

Dale, A., Arber, S. & Procter, M. 1988, *Doing Secondary Analysis*. Unwin Hyman, London, UK.

Dandridge, T. C. 1976, Symbols at work: the types and functions of symbols in selected organizations. Unpublished doctoral dissertation. University of California, Los Angeles.

Davis, S. 1984, *Managing Corporate Culture*. Ballinger, Cambridge, MA.

Davis, S. M. 1987, *Future Perfect*. Addison-Wesley, Reading, MA.

Dawes, R. M. 1972, *Fundamentals of Attitude Measurement*. Wiley, New York.

De Witte, K. & van Muijen, J. J. 1999, "Organizational culture: critical questions for researchers and practitioners", *European Journal of Work and Organizational Psychology*, vol. 8, no. 4, pp. 583–595.

Deal, T. E. & Kennedy, A. A. 1982, *Corporate Cultures: The Rites and Rituals of Corporate Life*. Perseus Publishing, Cambridge, MA.

Deal, T. E. & Kennedy, A. A. 1983, "Culture: a new look through old lenses", *Journal of Applied Behavioral Science*, vol. 19, pp. 488–505.

Deal, T. E. & Kennedy, A. A. 2000, *The New Corporate Cultures: Revitalizing the Workplace after Downsizing, Mergers and Reengineering*. Perseus Books Group, Philadelphia, PA.

Deetz, S. 1996, "Describing differences in approaches to organization science: rethinking Burrell and Morgan and their legacy", *Organization Science*, vol. 7, no. 2, pp. 191–207.

Deming, W. E. 1986, *Out of the Crisis*. MIT Centre for Advanced Engineering Study, Cambridge, MA.

Denison, D. R. 1990, *Corporate Culture and Organizational Effectiveness*. Wiley, New York.

Denison, D. R. 1996, "What is the difference between organizational culture and organizational climate? A native's point of view on a decade of paradigm wars", *Academy of Management Review*, vol. 21, pp. 619–654.

Denison, D. R. 2000, "Organizational culture: can it be a key lever for driving organizational change?", in *The Handbook of Organizational Culture*, S. Cartwright & C. L. Cooper, eds. John Wiley & Sons, London, UK.

Denison, D. R. & Mishra, A. 1995, "Towards a theory of organizational culture and effectiveness", *Organization Science*, vol. 6, no. 2, pp. 204–223.

Denison, D. R. & Mishra, A. O. 1998 Does organizational culture have an impact on quality? A study of culture and quality in ninety-two manufacturing organizations. Working Paper, University of Michigan.

Denison, D. R. & Neale, W. 1994, *Denison Organizational Culture Survey*. Aviat, Ann Arbor, MI.

Denison, D. R. and Neale, W. 1996 *Bringing Organizational Culture to the Bottom Line*. Available online HTTP: <http://www.denisonconsulting.com/dc/Portals/0/Docs/Paper_bottom_line.pdf>(accessed 5 July 2009).

Denison, D. R., Cho, H. J. & Young, J. 2000, Diagnosing organizational cultures: validating a model and method. Working Paper series no. 2000–9 in the collection of International Institute for Management Development, Laussane, Switzerland.

Denison, D. R., Hart, S. L. & Kahn, J. A. 1996, "From chimneys to cross-functional teams: developing and validating a diagnostic model", *Academy of Management Review*, vol. 39, pp. 1005–1023.

Denison Consulting. 2009, *The Denison Organizational Culture Survey*. Available online HTTP: <http://www.denisonconsulting.com/dc/Products/CultureProducts/CultureSurvey/tabid/40/Default.aspx> (accessed 29 August 2009).

Denison, D. R., Hoojiberg, R. & Quinn, R. E. 1995, "Paradox and performance: toward a theory of behavioral complexity in managerial leadership", *Organizational Science*, vol. 6, no. 5, pp. 524–540.

Denison, D. R., Lief, C. and Ward, J.L. 2004, 'Culture in family-owned enterprises: recognizing and leveraging unique strengths', *Family Business Review*, vol. 17, no. 1, pp. 61–70.

Denzin, N. K. 1989, *The Research Act*. McGraw-Hill, New York.

Denzin, N. K. & Lincoln, Y. S. 1994, *Handbook of Qualitative Research*. Sage, Thousand Oaks, CA.

Denzin, N. K. & Lincoln, Y. S. 1998, *Strategies of Qualitative Enquiry*. Sage, Thousand Oaks, CA.

DETR. 2000, *KPI Report for The Minister for Construction*. Department of Environment, Transport and the Regions (DETR), London, UK.

Donaldson, T. & Preston, L. E. 1995, "The stakeholder theory of corporation: concepts, evidence and implications", *Academy of Management Review*, vol. 20, pp. 65–91.

Doyle, P. 1994, "Setting business objectives and measuring performance", *Journal of General Management*, vol. 20, pp. 1–19.

Drucker, P. E. 1954, *The Practice of Management*. Harper & Row, New York.

Drucker, P. E. 1994, "The theory of business", *Harvard Business Review*, pp. 95–104.

Druckman, D., Singer, J. E., Van Cott, H. P., National Research Council (US), & Committee on Techniques for the Enhancement of Human Performance. 1997, *Enhancing organizational performance* National Academy Press, Washington, D.C.

DSS Research. 2004, *A Roadmap for Coming Health Care Market*. Available online HTTP: < http://www.dssresearch.com/home.asp> (accessed 5 July 2009).

DTI. 2002, *Quality Mark Scheme Member's Guide. The register of approved builders and tradesmen working in the domestic repair, maintenance and improvement sector*. Department of Trade and Industry, London, UK.

DTI. 2003, *Partners in Innovation 2002 - Call for Proposals & Guidelines for Applicants*. Department of Trade and Industry, London, UK.

Dulaimi, M. F., Ling, F. Y. Y., Ofori, G. and De Silva, N. 2002, "Enhancing integration and innovation in construction", *Building Research and Information*, vol. 30; no. 4, pp. 237–247.

Duncan, R. & Weiss, A. 1979, "Organizational learning: implications for organizational design", in *Research in Organizational Behavior, vol. 1*, B. M. Staw, ed. JAI Press, Greenwich, CT, pp. 75–123.

Easterby-Smith, M., Thorpe, R. & Lowe, A. 2002, *Management Research*, 2nd edn. Sage, London.

ECI Benchmarking Steering Committee. 1999, *Summaries of ECI Value Enhancement Practices*. European Construction Institute; Construction Industry Institute.

Eicher, L. D. 1992, ISO 9000 and related issues. Speech presented to an international conference *'Linking Standards to Practice'*.

Eitzen, D. S. & Yetman, N. R. 1972, "Managerial change, longevity and organizational effectiveness", *Administrative Science Quarterly*, vol. 17, pp. 110–116.

Eldrige, J. E. T. & Crombie, A. D. 1974, *A Sociology of Organizations*. Allen & Unwin, London, UK.

Ember, C. R. & Ember, M. 2001, *Cultural Anthropology*, 10th edn. Prentice Hall, Englowood Cliffs, NJ.

Encyclopædia Britannica. 2002, *Britannica Online 2002*. Available online HTTP: http://www.britannica.com/ (accessed 23 September 2002).

Enz, C. 1988, "The role of value congruity in intraorganizational power", *Administrative Science Quarterly*, vol. 33, pp. 284–304.

Etzioni, A. 1960, "Two approaches to organizational analysis: a critique and a suggestion", *Administrative Science Quarterly*, vol. 5, pp. 257–258.

European Foundation for Quality Management. *Introducing Excellence*. Available online HTTP: <http://excellenceone.efqm.org/Default.aspx?tabid=396 (accessed 1 December 2003).

Ewing, P. & Lundahl, L. 1996, The Balanced Scorecards at ABB Sweden - The EVITA Projects. EFI Research Paper 6567. Stockholm School of Economics, Stockholm, Sweden.

Faux, V. 1982, Unobtrusive controls in organizations: an action research approach to change. Unpublished doctoral dissertation, Harvard University, Cambridge, MA.

Fellows, R. & Liu, A. M. M. 2003, *Research Methods for Construction*, 2nd edn. Blackwell Publishing, Malden, MA.

Fellows, R. F. & Seymour, D. E. 2002, *Preface to Perspectives on Culture in Construction*. International Council for Research And Innovation in Building and Construction, Rotterdam, the Netherlands, CIB Publication no. 275.

Fey, C. F. & Denison, D. R. 2000, Organizational culture and effectiveness: the case of foreign firms in Russia. SSE/EFI Working Paper Series in Business Administration no. 2000: 4. Stockholm, Sweden.

Fey, C. F. & Denison, D. R. 2003, Organizational culture and effectiveness: can American theory be applied in Russia. William Davidson Institute Working Paper no. 598. Ann Arbor, MI.

Fisher, C. J. 2000, Like it or not, culture matters - linking culture to bottom line business performance', *Employee Relations Today*. vol. 27, no. 2.

Fisher, C. J. & Alford, R. J. 2000, "Consulting on culture: a new bottom line", *Consulting Psychology Journal*, vol. 52, summer, pp. 3–4.

Fitzgerald, L., Johnston, R., Brignall, S., Silvestro, R. & Voss, C. 1991, *Performance Measurement in Service Businesses*. CIMA, London, UK.

Folkman, J. 1998, *Employee Surveys That Make a Difference: Using Customized Feedback Tools to Transform your Organization*. Executive Excellence, Provo, UT.

Fombrun, C. 1983, "Corporate culture, environment, and strategy", *Human Resource Management*, vol. 22, pp. 139–152.

Forbes, L. H. 2001, Continuous improvement in the construction industry, *Leadership and Management in Engineering*, Winter, pp. 54–55.

Frost, P. J., Daft, R. L. & Lewin, A. 2000, *Talking About Organization Science: Debates & Dialogues from Crossroads*. Sage, Newbury Park, CA.

Frost, P. J., Moore, L. F., Louis, M. R., Lundberg, C. C. & Martin, J. 1991, *Reframing Organisational Culture*. Sage, London, UK.

Gable, G. G. 1994, "Integrating case study and survey research: an example in information systems", *European Journal of Information Systems*, vol. 3, no. 2, pp. 112–126.

Gabriel, Y. & Schwartz, H. S. 1999, "Individual and organization", in *Organizations in Depth: The Psychoanalysis of Organizations*, 1st edn, Y. Gabriel et al., eds. Sage, London, pp. 58–80.

Gagliardi, P. 1986, "The creation and change of organizational culture: a conceptual framework", *Organization Studies*, vol. 7, no. 2, pp. 117–134.

Gale, A. W. 1992, "The construction industry's male culture must feminize if conflict is to be reduced: the role of education as a gatekeeper to a male construction industry", in *Construction Conflict Management and Resolution*, P. Fenn & R. Gameson, eds. E and FM Spon, London, UK.

Garland, R. 1991, "The mid-point on a rating scale: is it desirable?", *Marketing Bulletin*, vol. 2, pp. 66–70.

Garrity, T. 2003, *Culture Essay*. Capistrano Valley High School. Article deleted but website is available online HTTP: <http://www.cvhs.com/cms/page_view?d=x&piid=&vpid=1211910101761> (originally accessed January 2003).

Geertz, C. 1966, "Religion as a cultural system", in *Anthropological Approaches to the Study of Religion*, M. Banton, ed. Tavistock Press, London, UK.

Gehrig, G. 1992, "Sabbatical leave enhancing construction management education", Auburn University, ed. Alabama, USA.

Georgopoulos, B. S. 1986, *Organizational Structure, Problem Solving, and Effectiveness*. Institute for Social Research, Ann Arbor, MI.

Gittings, D. 2000. "Blunders coming home to roost", *South China Morning Post*, 23 January, p. 1.

Glick, W. H. 1985, "Conceptualizing and measuring organizational and psychological climate: pitfalls in multilevel research", *Academy of Management Review*, vol. 10, no. 3, pp. 601–606.

Glisson, C. & James, L. R. 2002, "The cross-level effects of culture and climate in human service teams", *Journal of Organizational Behaviour*, vol. 23, pp. 767–794.

Goffman, E. 1969, "The characteristics of total institutions", in *A Sociological Reader on Complex Organizations*. A. Etzioni, ed. Rinehart & Winston, New York, pp. 312–356.

Goodenough, W. 1971, "Culture, language and society", Addison-Wesley Modular Publications no. 7. Reading, MA.

Goodman, P. S. & Pennings, J. M. 1977, *New Perspectives on Organizational Effectiveness*, 1st edn. Jossey-Bass, San Francisco, CA.

Gordon, G. G. 1985, "The relationship of corporate culture to industry sector and corporate performance", in *Gaining Control of the Corporate Culture*, R. H. Kilmann, M. J. Saxton & R. Serpa, eds. Jossey-Bass, San Francisco, CA, pp. 103–125.

Gordon, G. G. & DiTomaso, N. 1992, "Predicting corporate performance from organizational culture", *Journal of Management Studies*, vol. 29, no. 6, pp. 783–798.

Gray, R. 1998, *Organizational culture: a review of the literature*. Unpublished paper. Ashcroft International Business School, Cambridge and Chelmsford.

Great Britain Department of the Environment, T. a. t. R. C. T. F. 2002, *Rethinking Construction*. Department of the Environment, Transport and Regions (DETR), London, UK.

Green, F. B. 1992, "Organizational assessment for quality performance teamwork", *Work Teams Newsletter*, vol. 2, no. 1.

Green, F. B. & Henderson, D. A. 2000, "Nine ways to evaluate the effectiveness of your team-based organization", *The Journal for Quality and Participation*, vol. 23, no. 2, pp. 36–39.

Greene, J. C. and Caracelli, V. J. 2003, 'Making paradigmatic sense of mixed methods practice', in *Handbook of Mixed Methods in Social and Behavioral Research*, A. Tashakkori and C. Teddlie, eds. Sage, Thousand Oaks, CA, pp. 91–110.

Greene, J. C., Caracelli, V. J. & Graham, W. F. 1989, "Toward a conceptual framework for mixed-method evaluation designs", *Educational Evaluation and Policy Analysis*, vol. 2, pp. 255–274.

Gregory, K. L. 1983, "Native view paradigms: multiple cultures and culture conflicts in organizations", *Administrative Science Quarterly*, vol. 28, pp. 359–376.

Grinyer, P. H. & Norburn, D. 1975, "Planning for existing markets: perceptions of executives and financial performance", *Journal of the Royal Statistical Society* (A), vol. 138, pp. 70–79.

Hair J. F., Anderson, R. E., Tatham R. L. and Black, W. C. (1998). *Multivariate Data Analysis*, 5th edn. Prentice-Hall, Englewood Cliffs, NJ.

Hall, M. A. & Jaggar, D. M. 1997, "Accommodating cultural differences in international construction project procurement arrangements", CIB W92 (Procurement Systems) Conference, Montreal, Canada, May.

Hall, R. H. 1987, *Organizations: Structures, Processes and Outcomes*, 4th edn. Prentice-Hall., Englewood Cliffs, NJ.

Hallam, G. & Campbell, D. 1997, "The measurement of team performance with a standardized survey", in *Team Performance Assessment and Measurement: Theory, Methods and Applications*, M. T. Brannick, E. Salas & C. Prince, eds. Lawrence Erlbaum Associates, Inc., Mahwah, NJ., pp. 155–172.

Halpin, D. W. & Woodhead, R. W. 1998, *Construction Management*. John Wiley & Sons, New York.

Hamel, G. & Prahalad, C. K. 1989, "Strategic intent", *Harvard Business Review*, vol. 67, no. 3, pp. 63–76.

Hamel, G. & Prahalad, C. K. 1994, *Competing for the Future*. Harvard Business School Press, Boston, MA.

Handa, V. & Adas, A. 1996, "Predicting the level of organizational effectiveness: a methodology for the construction firm", *Construction Management and Economics*, vol. 14, pp. 341–352.

Handy, C. 1993, *Understanding Organizations*, 4th edn. Penguin, London, UK.

Hannan, M. T. & Freeman, J. 1977, "The population ecology of organizations", *American Journal of Sociology*, vol. 82, pp. 929–964.

Harris, M. 2004, *Culture, People, Nature: An Introduction to Anthropology*, 7th edn. Allyn & Bacon, Boston, MA.

Harrison, R. 1972, "Understanding your organization's character", *Harvard Business Review*, vol. 50, no. 23, pp. 119–128.

Harry, M. & Schroeder, R. 2000, *Six Sigma: The Breakthrough Management Strategy Revolutionizing the World's Top Corporations*. Random House Publishers, New York.

Hassard, J. & Pym, D. 1990, *The Theory and Philosophy of Organizations: Critical Issues and New Perspectives*. Routledge, London, UK.

Hatch, M. J. 2003, *Organization Theory: Modern, Symbolic and Postmodern Perspectives*. Oxford University Press, Oxford, UK.

Haviland, W. A. 1993, *A Cultural Anthropology*. Harcourt-Brace (Jovanovich College Publishers), Fort Worth, TX.

Hawking, S. W. 1998, *A Brief History of Time*, 10th Anniversary edn. Bantam Doubleday Dell, New York.

Hawkins, P. 1997, "Organizational culture: sailing between evangelism and complexity", *Human Relations*, vol. 50, no. 4, pp. 417–440.

Herskovits, M. J. 1948, *Man and His Works: The Science of Cultural Anthropology*. Knopf, New York.

Hirsch, P. M. 1975, "Organizational effectiveness and the institutional environment", *Administrative Science Quarterly*, vol. 20, pp. 327–344.

Hitt, M. A. 1988, "The measuring of organizational effectiveness: multiple domains and constituencies", *Management International Review*, vol. 28, no. 2, pp. 28–40.

Hitt, M. A. & Ireland, R. A. 1984, "Corporate distinctive competence and performance: effects of perceived environmental uncertainty, size and technology", *Decision Sciences*, vol. 15, no. 3, pp. 324–349.

Hitt, M. A. & Ireland, R. A. 1987, "Peters and Waterman revisited: the unended quest for excellence", *Academy of Management Executive*, vol. 1, no. 2, pp. 91–98.

Hitt, M. A., Ireland, R. A. & Stadter, G. A. 1982, "Functional importance and company performance: moderating effects of grand strategy and industry type", *Strategic Management Journal*, vol. 3, pp. 315–330.

Hofstede, G. H. 1991, *Cultures and Organizations Software of the Mind*. McGraw-Hill, London, UK.

Hofstede, G. & Bond, M. H. 1988, "The Confucius connection: from cultural roots to economic growth", *Organizational Dynamics*, vol. 16, no. 4, pp. 5–21.

Hofstede, G., Neuijen, B., Ohayv, D. D. & Sanders, G. 1990, "Measuring organizational cultures: a qualitative and quantitative study across twenty cases", *Administrative Science Quarterly*, vol. 35, pp. 286–316.

Hollway, W. 1991, *Work Psychology and Organisational Behaviour, Managing the Individual at Work*. Sage, London, UK.

Holsti, O. R. 1969, *Content Analysis for the Social Sciences and Humanities*. Addison-Wesley, Reading, MA.

Holt, G. D. & Proverbs, G. 2001, "A survey of public sector procurement in England", *Journal of Construction Procurement*, vol. 7, no. 1, pp. 3–10.

Hong Kong Housing Authority. 2000, *Quality Housing: Partnering for Change: A Consulting Document*. Hong Kong Housing Department, Hong Kong.

Hong Kong Housing Department. 1997, Performance Assessment Scoring System (PASS). Housing Department PASS Control Unit. Hong Kong Housing Department, Hong Kong.

Hong Kong Housing Department. 1999, *Information Paper for the LegCo Panel on Housing - Building Quality Assurance for Public Housing*. Hong Kong Housing Department, Hong Kong.

Howe, K. R. 2003, "Against the quantitative-qualitative incompatability thesis or dogmas die hard", *Educational Researcher*, vol. 17, pp. 10–16.

Hox, J. J. & Bechger, T. M. 1998, "An introduction to structural equation modelling", *Family Science Review*, vol. 11, pp. 354–373.

Hoy, W. K. & Miskel, C. G. 1996, *Educational Administration*, 5th edn. Random House, New York.

Hrebiniak, L. G. 1978, *Complex Organizations*. West Publishing, St. Pauls, MN.

Huczynski, A. & Buchanan, D. 1991, *Organizational Behaviour*, 2nd edn. Prentice-Hall, London, UK.

Human Synergistics International. 2004, *Organizational Culture Inventory (OCI)*. Available online HTTP: <http://www.human-synergistics.com.au/content/products/diagnostics/oci.asp> (accessed 5 July 2009).

Hunter, D. E. & Whitten, P. 1976, *Encyclopaedia of Anthropology*. Harper & Row, New York.

Integrated Publishing. *The Disadvantages of MBO*. Available online HTTP: <http://www.tpub.com/content/advancement/14144/css/14144_58.htm> (accessed 5 July 2009).

Jackson, B. 1999, *Perceptions of Organizational Effectiveness in Community and Member Based Nonprofit Organizations*. Doctoral Dissertation, University of La Verne, ProQuest Digital Dissertations, AAT9918843.

Jaeger, A. M. 1980, "The transfer of organizational culture overseas: an approach of control in the multinational corporation", *Journal of International Business Studies*, vol. 14, no. 2, pp. 91–114.

Janis, L. 1972, *Victims of Groupthink*. Houghton Mifflin, Boston, MA.

Jaques, E. & Tavistock Institute of Human Relations. 1951, *The Changing Culture of a Factory*. Tavistock Publications, London.

Jeffcutt, P. 2004, *The Foundations of Management Knowledge*. Routledge, New York.

Johnson, H. T. & Kaplan, R. S. 1987, *Relevance Lost: The Rise and Fall of Management Accounting*. Harvard Business School Press, Cambridge, MA.

Jones, G. 1983, "Transaction costs, property rights, and organizational culture: an exchange culture", *Administrative Science Quarterly*, vol. 28, pp. 454–467.

Jurmain, R., Nelson, H., Kilgore, L, & Trevathan, W. 2000, *Essentials of Physical Anthropology*, 5th edn. West Publishing, St Pauls, MN.

Kachigan, S. K. 1990, *Statistical Analysis: An Interdisciplinary Introduction to Univariate and Multivariate Methods*. Radius Press, New York, cha.15.

Kanter, R. 1983, *The Change Masters*. Simon & Schuster, New York.

Kaplan, B. & Duchon, D. 1988, "Combining qualitative and quantitative methods in information systems research", *MIS Quarterly*, vol. 12, no. 4, pp. 571–586.

Kaplan, R. S. & Norton, D. P. 1992, "The balanced scorecard: measures that drive performance", *Harvard Business Review*, vol. 70, pp. 71–79.

Kaplan, R. S. & Norton, D. P. 1996, *The Balanced Scorecard: Translating Strategy into Action*. Harvard Business School Press, Boston, MA.

Karger, D. W. & Malik, Z. A. 1975, "Long-range planning and organizational performance", *Long Range Planning*, December, pp. 60–64.

Katz, D. & Kahn, R. L. 1978, *The Social Psychology of Organizations*. John Wiley & Sons, New York.

Keegan, D. P., Eiler, R. G. & Jones, C. R. 1989, "Are your performance measures obsolete?", *Management Accounting*, vol. 70, no. 12, pp. 45–50.

Keeley, M. 1978, "A social-justice approach to organizational evaluation", *Administrative Science Quarterly*, vol. 23, pp. 272–292.

Keeley, M. 1984, "Impartiality and participant-interest theories of organizational effectiveness", *Administrative Science Quarterly*, vol. 29, no. 1, pp. 1–25.

Kennerley, M. & Neely, A. 2002, "Performance measurement frameworks: a review", in *Business Performance Measurement: Theory and Practice*, 1st edn, A. Neely, ed. Cambridge University Press, Cambridge, UK, pp. 145–155.

Kessing, R. M. & Strathern, A. J. 1998, *Cultural Anthropology*. Harcourt-Brace College Publishers, Fort Worth, TX.

Ket de Vries, M. 1980, *Organizational Paradoxes*. Tavistock, New York.

Kier, F. J. 1997, *Ways to explore the replicability of multivariate results (since statistical significance testing does not)*. Paper presented at the annual meeting of the Southwest Educational Research Association, Austin, TX.

Kilmann, R. H., Saxton, M. J. & Serpa, R. 1985, *Gaining Control of the Corporate Culture*, 1st edn. Jossey-Bass, San Francisco, CA.

Kimberly, J. R. & Nielsen, W. R. 1975, "Organization development and change in organizational performance", *Administrative Science Quarterly*, vol. 20, pp. 191–206.

Kirchoff, B. A. 1977, "Organization effectiveness measurement and policy research", *Academy of Management Review*, vol. 2, no. 3, pp. 347–355.

Kluckhohn, C. K. M. & Kelly, W. H. 1945, "The concept of culture. From *The Science of Man in World Crisis*", in *Culture and Behavior*, C. K. M. Kluckhohn, ed. Free Press, New York, pp. 20–73.

Koene, B. A. S. 1996, Organizational culture, leadership and performance in context: trust and rationality in organizations. Unpublished doctoral dissertation, Rijkuniversiteit Limburg.

Kotelnikov, V. 2004, *Management By Objectives*. Available online HTTP: <http://www.1000ventures.com/business_guide/mgmt_mbo_main.html> (accessed 5 July 2009).

Kottak, C. P. 1994, *Cultural Anthropology*, 6th. Edn. McGraw-Hill, New York.

Kotter, J. P. & Heskett, J. L. 1992, *Corporate Culture and Performance*. Free Press, New York.

Kreitner, R. & Kinicki, A. 2001, *Organizational Behavior*, 5th edn. McGraw-Hill, London, UK.

Krippendorff, K. 1980, *Content Analysis: An Introduction to its Methodology*. Sage, Newbury Park, CA.

Kroeber, A. L. 1948, *Anthropology*. Harcourt, Brace & Jovanovich, New York.

Kroeber, A. L. & Kluckhohn, C. K. M. 1952, Culture: a critical review of concepts and definitions [47], 1. Papers of the Peabody Museum of Archaeology and Ethnology.

Kumaraswamy, M. M. & Dissanayaka, S. M. 1996, "Procurement by objectives", *Journal of Construction Procurement*, vol. 2, no. 2, pp. 38–51.

Kunda, G. 1992, *Engineering Culture: Control and Commitment in a High-tech Corporation*. Temple University Press, PA.

Lafferty, J. C. 1973, *Leadership and Motivation*. Human Synergistics, Plymouth, MI.

Lambert, B. H., Funato, K. & Poor, A. 1996, The construction industry in Japan and its subcontracting relationships (EIJS WP no. 21). Working Paper for the European Institute of Japanese Studies, Stockholm School of Economics.

Langetieg, T. 1978, "An application of a three-factor performance index to measure stockholder gains from merger", *Journal of Financial Economics*, vol. 6, pp. 365–383.

Langford, D. A. and Fellows, R. N. R. 1990a, *Construction Management 1 - Management Systems*. The Mitchell Publishing Company, London, UK.

Langford, D. A. & Fellows, R. N. R. 1990b, *Construction Management 2 - Organisation Systems*. The Mitchell Publishing Company, London, UK.

Lansley, P. 1994, "Analysing construction organizations", *Construction Management and Economics*, vol. 12, pp. 337–348.

Latham, S. M. 1994, *Constructing the Team*. Her Majesty's Stationery Office (HMSO), London, UK.

Lawler, E. E. 1977, "The new plant revolution", *Organization Dynamics*, vol. 6, no. 3, pp. 2–12.

Lawler, E. E. 1986, *High Involvement Management: Participative Strategies for Improving Organizational Performances*. Jossey-Bass, San Francisco, CA.

Lawler, E. E., Nadler, D. & Cammann, C. 1980, *Organizational Assessment Perspectives on the Measurement of Organizational Behavior and the Quality of Work Life*. Wiley, New York.

Lawrence, P. R. & Lorsch, J. W. 1967, *Organization and Environment*. Harvard Business School, Cambridge, MA.

Leavitt, H. J. 1965, "Applied organizational change in industry: structural, technological and human approaches", in *Handbook of Organizations*, J. G. March, ed. Rand McNally, Chicago, ILL, pp. 1144–1170.

Lee, A. S. 1991, "Integrating positivist and interpretive approaches to organizational research", *Organization Science*, vol. 2, pp. 342–365.

Leung, M. Y. 1999, *From Shelter to Home: 45 Years of Public Housing Development in Hong Kong*. Hong Kong Housing Authority, Hong Kong.

Lévi-Strauss, C. 1969, *The Elementary Structures of Kinship (Les Structures élémentaires de la Parenté)*, revised edition translated from the original French version (1949) by James Harle Bell; edited by John Richard von Sturmer and Rodney Needham. Eyre & Spottiswoode, London, UK.

Lewin, K. 1951, "Field theory in social science: selected theoretical papers", D. Cartwright, ed. Harper & Row, New York.

Lewin, K., Lippitt, R. & White, R. K. 1939, "Patterns of aggressive behavior in experimentally created social climates", *Journal of Social Psychology*, vol. 10, pp. 271–279.

Lewis, D. 1996, "The organizational culture saga - from OD to TQM: a critical review of the literature. Part 1 - concepts and early trends", *Leadership and Organization Development Journal*, vol. 17, no. 1, pp. 12–19.

Lewis, D. S. 1994, "Organizational change: relationship between reactions, behaviour and organizational performance", *Journal of Organizational Change Management*, vol. 7, no. 5, pp. 41–55.

Likert, R. 1967, *The Human Organization: Its Management and Value.* McGraw-Hill, New York.

Likert, R. L. 1961, *New Patterns of Management.* McGraw-Hill, New York.

Lim, B. 1995, "Examining the organizational culture and organizational performance link", *Leadership and Organizational Development Journal*, vol. 16, no. 5, pp. 16–21.

Lincoln, Y. S. & Guba, E. G. 1985, *Naturalistic Inquiry.* Sage, Beverly Hills, CA.

Linderman, K., Schroeder, R. G., Zaheer, S. & Choo, A. S. 2003, "Six sigma: a goal-theoretic perspective", *Journal of Operations Management*, vol. 21, pp. 193–203.

Lingard, H. & Rowlinson, S. M. 1998, "Behaviour-based safety management in Hong Kong's construction industry: the results of a field study", *Construction Management and Economics*, vol. 16, no. 4, pp. 481–488.

Lingle, J. H. & Schieman, W. A. 1996, "From balanced scorecard to strategic gauges: is measurement worth it?", *Management Review*. March, pp. 56–61.

Linton, R. 1945, *The Cultural Background of Personality.* Appleton-Century, New York.

Litwin, G. H. & Stringer, R. 1968, *Motivation and Organizational Climate.* Harvard University Press, Cambridge, MA.

Liu, A. M. M. 1999, "Culture in the Hong Kong real-estate profession: a trait approach", *Habitat International*, vol. 23, no. 3, pp. 413–425.

Liu, A. M. M. 2002, *Keys to Harmony and Harmonic Keys.* International Council for Research and Innovation in Building and Construction, Rotterdam, the Netherlands, CIB TG23 Culture in Construction Publication 275.

Liu, A. M. M. 2003, "Organizational culture profiles of the Chinese contractors", Department of Real Estate and Construction, The University of Hong Kong, Hong Kong, CIB TG 23 International Conference, pp. 1–10.

Liu, A. M. M. & Fellows, R. F. 1999a, "Cultural issues", in *Procurement Systems: A Guide to Best Practice in Construction*, S. M. Rowlinson & P. McDermott, eds. E and FM Spon, London, UK, pp. 141–162.

Liu, A. M. M. & Fellows, R. F. 1999b, "The impact of culture on project goals", in *Profitable Partnering in Construction Procurement*, S. Ogunlana, ed. E and FM Spon, London, UK, pp. 523–532.

Locke, E. A. 1968, "Toward a theory of task performance and incentives", *Organizational Behavior and Human Decision Processes*, vol. 3, pp. 157–189.

Locke, E. A. & Schweiger, D. M. 1979, "Participation in decision-making: one more look", *Research in Organizational Behavior*, vol. 1, pp. 265–339.

Lorsch, J. 1986, "Managing culture: the invisible barrier to strategic change", *California Management Review*, vol. 28, pp. 95–109.

Loubser, J. 2003, *Definitions of Culture.* Salford University, UK. Abstract of paper deleted but website is available online HTTP: <http://www.business.salford.ac.uk/modules/module2.php?module=521> (accessed December 2003).

Louis, M. R. 1985, "An investigator's guide to workplace culture", in *Organizational Culture*, P. J. Frost et al., eds. Sage, Beverley Hills, CA, pp. 73–94.

Love, P. E. D., Li, H., Irani, Z. & Holt, G. D. 2000, "Re-thinking TQM: toward a framework for facilitating learning and change in construction organisations", *TQM Magazine*, vol. 12, no. 2, pp. 107–116.

Low, S. P. & Yeo, H. K. C. 1998, "A construction quality costs qualifying system for the building industry", *International Journal of Quality and Reliability Management*, vol. 15, no. 3, pp. 329–349.

Lubatkin, M. 1983, "Mergers and performance of the acquiring firm", *Academy of Management Journal*, vol. 8, no. 2, pp. 218–225.

Lynch, R. L. & Cross, K. F. 1991, *Measure Up! The Essential Guide to Measuring Business Performance.* Mandarin, London, UK.

M4I Board 2000, *Movement for Innovation.* Movement for Innovation (Building Research Establishment), Garstang, UK.

Maisel, C. S. 1992, "Performance measurement: the balanced scorecard approach", *Journal of Cost Management,* Fall, pp. 47–52.

Maital, S. 1984, "The economic analysis of the Japanese firm", *Sloan Business Review,* vol. 26, no. 1, pp. 77–78.

Malinowski, B. 1944, *A Scientific Theory of Culture and Other Essays.* University of North Carolina Press, Chapel Hill, NC.

Maloney, W. F. & Federle, M. O. 1991, "Organizational culture and management", *Journal of Management and Engineering,* vol. 7, no. 1, pp. 1–25.

Maraglia, E., Law, R. & Collins, P. 1996. *Culture Index.* Available online HTTP: http://www.wsu.edu:8001/vcwsu/commons/topics/culture/culture-index.html> (accessed 23 January 2003).

Marcoulides, G. A. & Heck, R. H. 1993, "Organizational culture and performance: proposing and testing a model", *Organization Science,* vol. 4, pp. 209–225.

Marett, R. R. 1936, *Modern Sociologists – Tylor.* John Wiley & Sons, New York.

Martin, J. 1992, *Cultures in Organizations.* Oxford University Press, New York.

Martin, J. 2002, *Organizational Culture - Mapping the Terrain.* Sage, Thousand Oaks, CA.

Martin, J. & Frost, P. J. 1996, "The organizational culture war games: a struggle for intellectual dominance", in *Handbook of Organization Studies,* S. R. Clegg, C. Hardy & W. R. Nord, eds. Sage, London, UK.

Martin, J. & Meyerson, D. E. 1988, "Organizational cultures and the denial, channeling and acknowledgement of ambiguity", in *Managing Ambiguity and Change,* L. Pondy, R. Boland & H. Thomas, eds. Wiley, New York, pp. 93–125.

Martin, J. & Siehl, C. 1983, "Organizational culture and counter culture", *Organizational Dynamics,* vol. 12, pp. 52–64.

Martin, J. R. 1978, A summary of Ouchi, W. G. and A. M. Jaeger. 1978. Type Z Organization: stability in the midst of mobility, Academy of Management Review April, pp. 305–314.

Martinez-Lorente, A. R., Dewhurst, F. & Dale, B. G. 1998, 'Total quality management: origins and evolution of the term', *TQM Magazine,* vol. 10, no. 5, pp. 378–386.

McCarthy, P. M., Fleishman, E. A. & Holt, R. 1997. "Where's the evidence that managers' long-term performance really does synchronize with established organizational performance dimensions?", *Institute of Behavioral and Applied Management Proceedings,* vol. 5, pp. 85–91.

McElroy, D. *Biology Course #483, Multivariate Methods in Biology.* Western Kentucky University. Available online HTTP: <http://bioweb.wku.edu/courses/courses.asp#483> (accessed 5 July 2009).

McGee, R. J. and Warms, R .L. 2004, *Anthropological Theory: An Introductory History,* McGraw Hill, New York.

McGregor, D. 1960, *The Human Side of Enterprise.* McGraw-Hill, New York.

McKiernan, P. & Morris, C. 1994, "Strategic planning and financial performance in UK SMEs: does formality matter?", *British Journal of Management,* vol. 10, pp. 75–87.

McNamara, C. 2003, *Basic Definition of Organization.* Minnesota Organization Development Network (MNODN). Available online HTTP: <http://managementhelp.org/org_thry/org_defn.htm> (accessed 5 July 2009).

Mead, G. H. 1934, *Mind, Self and Society.* Chicago University Press, Chicago, ILL.

Mead, M. 1937, *Cooperation and Competition Among Primitive Peoples.* McGraw-Hill, London, UK.

Meek, V. L. 1988, "Organizational culture: origins and weaknesses", *Organization Studies*, vol. 9, no. 4, pp. 453–473.

Mercury Group, Donahue, K. 2002, *Fish See the Water Last: Organizational Effectiveness*. Available online (subscription required) HTTP: <http://store.astd.org/Default.aspx?tabid=143&action=ECDProductDetails&args=16918> (home page accessed 5 July 2009).

Merriam-Webster Incorporated. 2002, *Merriam-Webster Online Dictionary*. Available online HTTP: <http://www.merriam-webster.com/> (accessed 5 July 2009).

Meyerson, D. E. 1991, "Acknowledging and uncovering ambiguities in cultures", in *Reframing Organizational Culture*, P. J. Frost et al., eds. Sage, Newbury Park, CA, pp. 254–270.

Microsoft Corporation. 2002, *Encarta World English Dictionary* [North American Edition]. Available online HTTP: <http://encarta.msn.com/encnet/features/dictionary/dictionaryhome.aspx> (accessed 5 July 2009).

Miles, J. & Ezzell, J. 1980, "The weighted average cost of capital, perfect capital markets, and project life: clarification", *Journal of Financial and Quantitative Analysis*, vol. 15, September, pp. 719–730.

Miller, D. & Mintzberg, H. 1983, "The case for configuration", in *Beyond Method: Strategies for Social Research*, G. Morgan, ed. Sage, Beverly Hills, CA, pp. 57–73.

Miller, K. I. & Monge, P. R. 1986, "Participation, satisfaction, and productivity: a meta-analytic review", *Academy of Management Review*, vol. 29, pp. 727–753.

Mintzberg, H. 1987, "Crafting strategy", *Harvard Business Review*, vol. 65, pp. 66–75.

Mintzberg, H. 1994, *The Rise and Fall of Strategic Planning: Reconciling for Planning, Plans, Planners*. Free Press, New York.

Moch, M. & Seashore, S. E. 1981, "How norms affect behaviors in and of corporations", in *Handbook of Organizational Design, Vol. 1*, P. Nystrom & W. Starbuck, eds. Oxford University Press, New York, pp. 210–237.

Mohamed, S. 2003, "Performance in international construction joint ventures: modelling perspective", *Journal of Construction Engineering and Management*, vol. 129, no. 6, pp. 619–626.

Mohan, L. 1993, *Organizational Communication and Cultural Vision*. State University of New York Press, New York.

Mohr, L. B. 1983, "The reliability of the case study as a source of information", in *Symposium on Information Processing in Organizations*, R. Coulam & R. Smith, eds, JAI Press.

Molnar, J. J. & Rogers, D. L. 1976, "Organizational effectiveness: an empirical comparison of the goal and system resources", *Sociological Quarterly*, vol. 17, pp. 401–413.

Moore, L. F. 1985, "How are organizational cultures and the wider cultural context linked?", in *Organizational Culture*, P. J. Frost et al., eds. Sage, Beverly Hills, CA, pp. 277–280.

Morgan, D. L. & Krueger, R. A. 1993, "When to use focus groups and why", in *Successful Focus Groups: Advancing the State of the Art*, D. L. Morgan, ed. Sage, Newbury Park, CA, pp. 3–19.

Morris, M. W., Leung, K., Ames, D. & Lickel, B. 1999, "Views from inside and outside: integrating emic and etic insights about culture and justice", *Academy of Management Review*, vol. 24, no. 4, pp. 781–796.

Motulsky, H. 2004, *Analyzing data with GraphPad Prism*. Available online: <http://www.graphpad.com/articles/AnalyzingData.pdf> (accessed 5 July 2009).

Motulsky, H. and Christopoulos, A. 2003, *Fitting Models to Biological Data Using Linear and Nonlinear Regression: A Practical Guide to Curve Fitting*. Oxford University Press, New York.

Motwani, J. 1997, "Viewpoint: total quality management or totalled quality management", *International Journal of Quality and Reliability Management*, vol. 14, no. 7, pp. 647–650.

Motwani, J. 2001a, "Measuring critical factors of TQM", *Measuring Business Excellence*, vol. 5, no. 2, pp. 27–30.

Motwani, J. 2001b, "Critical factors and performance measures of total quality management", *TQM Magazine*, vol. 13, no. 4, pp. 292–300.

Murcko, T. 2004, *Investorguide.com*. WebFinance Inc. Original article no longer available but website Available online HTTP: <http://www.investorguide.com/> (accessed 5 July 2009).

Murdock, G. P., Ford, C. S., Hudson, A. E., Kennedy, R., Simmons, L. W. & Whiting, J. W. M. 1983, *Outline of Cultural Materials*, 5th rev. edn. Human Relations Area Files, New Haven, CT.

Murphy, D. M. D. 2003, *Definitions of Culture*. Department of Anthropology, University of Alabama. Article deleted but website Available online <http://web.as.ua.edu/ant/index.php> (accessed 5 July 2009).

Murphy, R. 1986, *Culture and Social Anthropology*. Prentice-Hall, Englewood Cliffs, NJ.

Neely, A. & Adams, C. 2001, "Perspectives on performance: the performance prism", *Journal of Cost Management*, vol. 15, no. 1, pp. 7–15.

Neely, A., Gregory, M. & Platts, K. 1995, "Performance measurement system design: a literature review and research agaenda", *International Journal of Operations and Production Management*, vol. 15, no. 4, pp. 80–116.

Newcombe, R., Langford, D. & Fellows, R. 1990, *Construction Management*. Mitchell, London, UK.

Nichols, J. 1984, "An alloplastic approach to corporate culture", *International Studies of Management and Organization*, vol. 14, pp. 32–63.

Norusis, M. J. 2002, *SPSS 11.0 Guide to Data Analysis*. Prentice Hall, London, UK.

Noyes, C. M. 1992, *Creativity, Change and Culture: An Investigation into the Relationship Between Organisational Culture and Innovation*. UMI Research Press, Ann Arbor, MI.

Nunally, J. C. 1970, *Introduction to Psychological Measurement*. McGraw-Hill., New York.

O'Brien, D. 2003, *Cultural Definitions*. Temple University, Philadelphia. Article deleted but website available online HTTP: <http://www.temple.edu/anthro/faculty.htm#obrien> (accessed 5 July 2009).

O'Donnell, N. H. & Matthews, J. M. 2001, *Construction Sub-contracting*. Federal Publications Seminars LLC, VA. Abstract deleted but first author's website available online HTTP: <http://www.rjo.com/pub_construction.html> (accessed 5 July 2009).

O'Reilly, C. A., III, Chatman, J. A. & Caldwell, D. F. 1991, "People and organizational culture: a profile comparison approach to assessing person-organization fit", *Academy of Management Journal*, vol. 34, pp. 487–516.

Odiorne, G. S. 1965, *Management by Objectives: A System of Managerial Leadership*. Pitman, New York.

Ohmae, K. 1982, *The Mind of the Strategist: The Art of Japanese Business*. McGraw-Hill, New York.

Ortner, S. B. 1984, "Theory in anthropology since the sixties", *Comparative Studies in Society and History*, vol. 26, pp. 126–166.

Ostroff, C. & Schmitt, N. 1993, "Configurations of organizational effectiveness and efficiency", *Academy of Management Journal*, vol. 36, no. 6, pp. 1175–1195.

Ott, J. 1989, *The Organizational Culture Perspective*. Brooks and Cole, Pacific Grove, CA.

Ouchi, W. G. 1981, *Theory Z: How American Business Can Meet the Japanese Challenge*. Addison-Wesley, Reading, MA.

Ouchi, W. G. & Jaeger, A. M. 1978, "Type Z organization: stability in the midst of mobility", *Academy of Management Review*, vol. 3, no. 2, pp. 305–314.

Ouchi, W. G. & Johnson, J. B. 1978, "Types of organizational control and their relationship to emotional well-being", *Administrative Science Quarterly*, vol. 23, no. 2, pp. 293–317.

Ouchi, W. G. & Price, R. 1978, "Hierarchies, clans, and Theory Z: a new perspective on organizational development", *Organizational Dynamics*, vol. 7, no. 2, pp. 25–44.

Pacanowsky, M. E. & O'Donnell-Trujillo, N. 1982, "Communication and organizational culture", *The Western Journal of Speech Communication*, vol. 46, spring, pp. 115–130.

Paine, F. T. & Anderson, C. R. 1977, "Contingencies affecting strategy formulation and effectiveness: an empirical study", *Journal of Management Studies*, vol. 14, pp. 147–158.

Pande, P. S., Neuman, R. P. & Cavanagh, R. R. 2000, *The Six Sigma Way: How GE, Motorola and other Top Companies are Honing Their Performance*. McGraw-Hill Professional, New York.

Parsons, T. 1960, *Structure and Process in Modern Societies*. Free Press, Glencoe, ILL.

Pascale, R. T. 1984, "Fitting new employees into the company culture", *Fortune Magazine*, vol. 109, no. 11, pp. 28–43.

Pascale, R. T. 1985, "The paradox of "corporate culture": reconciling ourselves to socialization", *California Management Review*, vol. 27, no. 2, pp. 26–41.

Pascale, R. T. & Athos, A. G. 1981, *The Art of Japanese Management*. Simon & Schuster, New York.

Patton, M. Q. 1990, *Qualitative Evaluation and Research Methods*. Sage, Newbury Park, CA.

Payne, R. L. 2002, "Climate and culture: how close can they get?", in *Handbook of Organizational Culture and Climate*, 1st edn, N. M. Ashkanasy, C. Wilderom & M. F. Peterson, eds. Sage, Thousand Oaks, CA, pp. 163–176.

Pennings, J. M. 1976, "Dimensions of organizational influence and their effectiveness correlates", *Administrative Science Quarterly*, vol. 21, pp. 688–699.

Pennings, J. M. & Gresov, C. 1986, "Technoeconomic and structural correlates of organizational culture", *Organization Studies*, vol. 7, pp. 317–334.

Peoples, J. & Bailey, G. 1999, *Humanity: An Introduction to Cultural Anthropology*, 6th edn. West Publishing, St Pauls, MN.

Peters, T. J. & Austin, N. 1985, *A Passion for Excellence*. Random House, New York.

Peters, T. J. & Waterman, R. H. 1982, *In Search of Excellence: Lessons from America's Best-run Companies*. Harper & Row, New York.

Peters, T. J., Waterman, R. H. & Phillips, J. R. 1980, "Structure is not organization", *Business Horizons*, vol. 23, no. 3, pp. 14–21.

Petter, S. C. & Gallivan, M. J. 2004, "Toward a framework for classifying and guiding mixed method research in information systems", *Proceedings of the 37th Hawaii International Conference on System Sciences*, Waikiki, HI.

Pettigrew, A. 1979, "On studying organizational cultures", *Administrative Science Quarterly*, vol. 24, pp. 570–581.

Pettigrew, A. 1985, *The Awakening Giant*. Blackwell, Oxford, UK.

Petty, M. M. & Bruning, N. W. 1980, "Relationship between employees' attitudes and error rates in public welfare programs", *Academy of Management Journal*, vol. 23, pp. 556–561.

Petty, M. M., Beadles, N. A. I., Lowery, C. M., Chapman, D. F. & Connell, D. W. 1995, "Relationships between organizational culture and organizational performance", *Psychological Reports*, vol. 76, pp. 483–492.

Pfeffer, J. 1981, "Management as symbolic action: the creation and mainte-nance of organizational paradigms", in *Research in Organizational Behavior*, L. L. Cummings & B. M. Staw, eds. JAI Press, Greenwich, CT.

Pfeffer, J. & Salncik, G. R. 1978, *The External Control of Organizations: A Resource Dependence Perspective*. Harper & Row, New York.

Pfiffner, J. M. & Sherwood, F. P. 1960, *Administrative Organization*. Prentice Hall, Englewood Cliffs, NJ.

Pike, K. L. 1967, *Language in Relation to a Unified Theory of the Structure of Human Behavior*, 2nd edn. Mouton, The Hague, NL.

Pondy, L. R., Frost, P. J., Morgan, G. & Dandridge, T. C. 1983, *Organizational Symbolism*. JAI Press, Greenwich, CT.

Potter, C. C. 1989, "What is culture and can it be useful for organizational change agents?", *Leadership and Organization Development Journal*, vol. 10, no. 3, pp. 17–24.

Preston, L. E. & Spienza, H. J. 1990, "Stakeholder management and corporate performance", *Journal of Behavioral Economics*, vol. 19, pp. 361–375.

Price-Williams, D. 1974, "Psychological experiment and anthropology: the problem of categories", *Ethos*, vol. 2, pp. 95–114.

Product Development & Management Association. 2004, *Glossary of New Product Development Terms*. Available online HTTP: <http://www.pdma.org/npd_glossary.cfm> (accessed 5 July 2009).

Provisional Construction Industry Co-ordination Board. 2002, *Management of Sub-contracting in Construction*. HKSAR Government., Hong Kong. Paper no. PCICB/005.

Pun, K. F. 2002, "Development of an integrated TQM and performance measurement system for self-assessment: a method", *Total Quality Management*, vol. 13, no. 6, pp. 759–777.

Quinn, R. E. 1991, *Beyond Rational Management: Mastering the Paradoxes and Competing Demands of High Performance*, reprint edn. Jossey-Bass, San Francisco, CA.

Quinn, R. E. & Cameron, K. S. 1988, "Paradox and transformation", in *Paradox and Transformation*, R. E. Quinn & K. S. Cameron, eds. Ballinger, Cambridge, MA.

Quinn, R. E. & Rohrbaugh, J. 1983, "A spatial model of effectiveness criteria: towards a competing values approach to organizational analysis", *Management Science*, vol. 29, pp. 363–377.

Radcliffe-Brown, A. R. 1949, "White's view of a science of culture", *American Anthropologist*, vol. 51, pp. 503–512.

Rainey, H. G. 2003, *Understanding and Managing Public Organizations*, 3rd edn. John Wiley & Sons, London, UK.

Reber, A. S. ed. 1995, *Penguin Dictionary of Psychology*, 2nd edn. Penguin Books, London, UK.

Reichardt, C. S. & Rallis, S. F. 1994, "Qualitative and quantitative enquiries are not incompatible: a call for a new partnership", in *The Qualitative-Quantitative Debate: New Perspectives*, C. S. Reichardt & S. F. Rallis, eds. Jossey-Bass, San Francisco. CA, pp. 85–92.

Reichers, A. & Schneider, B. 1990, "Climate and culture: an evolution of constructs", in *Organizational Climate and Culture*, B. Schneider, ed. Jossey-Bass, San Francisco, CA, pp. 5–39.

Reilly, F. K. & Brown, K. 1979, *Investment Analysis and Portfolio Management*, 1st edn. The Dryden Press, Orlando, FL.

Rethinking Construction Ltd. 2002a, *Accelerating Change: A Report by the Strategic Forum for Construction*. Department of Trade and Industry, London, UK.

Rethinking Construction Ltd. 2002b, *Respect for People: A Framework for Action*. Department of Trade and Industry, London, UK.

Reynolds, P. D. 1986, "Organizational culture as related to industry, position and performance: a preliminary report", *Journal of Management Studies*, vol. 23, pp. 333–345.

Ridley, C. R. & Mendoza, D. W. 1993, "Putting organizational effectiveness into practice: the preeminent consultation task", *Journal of Counseling and Development*, vol. 72, pp. 168–178.

Robbins, S. 1998, *Organisational Behaviour: Concepts, Controversies and Applications*, 8th edn. Prentice Hall, Englewood Cliffs, NJ.

Robbins, S. P. 1990, *Organization Theory: Structure, Design and Applications*, 3rd edn. Prentice-Hall, Englewood Cliffs, NJ.

Robbins, S. R. & Duncan, R. B. 1988, "The role of the CEO and top management in the creation and implementation of strategic vision", in *The Executive Effect*, D. C. Hambrick, ed. JAI Press, Greenwich, CT.

Roethlisberger, F. J. & Dickson, W. J. 1939, *Management and the Worker: An Account of a Research Programme Conducted by Western Electric Company, Hawthorne Works Chicago*. Harvard University Press, Cambridge, MA.

Rojas, R. R. 2000, "A review of models for measuring organizational effectiveness among for-profit and nonprofit organizations", *Nonprofit Management and Leadership*, vol. 15, nos 1/2, pp. 97–104.

Rokeach, M. 1972, *Beliefs, Attitudes and Values: A Theory of Organization and Change*, 2nd edn. Jossey-Bass, San Francisco, CA.

Rollins, T. & Roberts, D. 1998, *Work Culture, Organizational Performance, and Business Success: Measurement and Management*. Quorum Books, Westport, CT.

Root, F.R. 1994, *Entry Strategies for International Markets*. Lexington Books, New York.

Root, D. S. 2001, The influence of professional and occupational cultures on project relationships mediated through standard forms and conditions of contract. Ph.D., Construction Management, University of Bath.

Root, D. S. 2004. *The Effect of Culture on Project Performance Under the New Engineering Contract*. ARCOM, University of Reading, UK.

Rosen, M. 1985, "Breakfast at Spiro's: dramaturgy and dominance", *Journal of Management*, vol. 11, pp. 31–48.

Rouncefield, P. 2004, *Sociology Stuff*. Work Teams Newsletter.

Rousseau, D. M. 1990, "Normative beliefs in fund-raising organizations: linking culture to organizational performance and individual responses", *Group and Organization Studies*, vol. 15, pp. 448–460.

Rowlinson, M. & Procter, S. 1999, "Organizational culture and business history", *Organization Studies*, vol. 20, no. 3, pp. 369–396.

Rowlinson, S. M. 2001, "Matrix organizational structure, culture and commitment: a Hong Kong public sector case study of change", *Construction Management and Economics*, vol. 19, pp. 669–673.

Rowlinson, S. M. & Lingard, H. 1998, "Behaviour modification - a new approach to improving site safety in Hong Kong", Technical Seminar Jointly Presented by the Building Division and the Asian Construction Management Association, Hong Kong Institute of Engineers, Hong Kong.

Rowlinson, S. M. & Root, D. S. 1996, *The Impact of Culture on Project Management - Final Report to the British Council*. Department of Real Estate and Construction, The University of Hong Kong, Hong Kong.

Rumelt, R. P. 1974, *Strategy, Structure, and Economic Performance*. Harvard Business Press, Cambridge, MA.

Sackmann, S. 1991, "Uncovering culture in organizations", *Journal of Applied Behavioural Science*, vol. 27, pp. 295–317.

Saffold, G. S. 1988, "Culture traits, strength and organizational performance: moving beyond strong culture", *Academy of Management Review*, vol. 13, no. 4, pp. 546–558.

Salazar, N. 1989, "Applying the deming philosophy to the safety system", *Professional Safety*, December, pp. 22–27.

Sashkin, M. & Fullmer, R. M. 1985, "Measuring organizational excellence", Conference paper presented at the national meeting of the Academy of Management, San Diego.

Sathe, V. 1983, "Implications of corporate culture: a manager's guide to action", *Organizational Dynamics*, vol. 12, no. 2, pp. 5–23.

Sathe, V. 1985, "How to decipher and change culture", in *Gaining Control of the Corporate Culture*, R. H. Kilmann, M. J. Saxton & R. Serpa, eds. Jossey-Bass, Oxford.

Saunders, M., Lewis, P. & Thornhill, A. 2000, *Research Methods for Business Students*, 2nd edn. FT Prentice Hall, Harlow, England.

Scandura, T. A. & Williams, E. A. 2000, "Research methodology in management: current practices, trends, and implications for future research", *Academy of Management Journal*, vol. 43, no. 6, pp. 1248–1265.

Schaffer, B. S. & Riordan, C. M. 2003a, "A review of cross-cultural methodologies for organizational research using self-report measures: a best practices approach", *Organizational Research Methods*, vol. 6, no. 2, pp. 169–216.

Schaffer, B. S. & Riordan, C. M. 2003b. Cross-cultural methodologies for organizational research using self-report measures: a best practices approach, *Organizational Research Methods*, vol. 6, no. 2, 169–215.

Schall, M. 1983, "A communications-rules approach to organizational culture", *Administrative Science Quarterly*, vol. 28, pp. 557–581.

Schauber, A. C. 2001, "Effecting extension organizational change toward cultural diversity: a conceptual framework", *Journal of Extension*, vol. 39, no. 3, pp. 1–8.

Schein, E. 1983, "The role of the founder in creating organizational culture", *Organizational Dynamics*, vol. 12, no. 1, pp. 13–28.

Schein, E. 1984, "Coming to a new awareness of organizational culture", *Sloan Management Review*, vol. 25, pp. 3–16.

Schein, E. 1985, *Organizational Culture and Leadership*. Jossey-Bass, San Francisco, CA.

Schmidt, F. L. & Hunter, J. E. 1977, "Development of a general solution to the problem of validity generalization", *Journal of Applied Psychology*, vol. 62, pp. 529–540.

Schneider, B. 1994, *Organizational Behaviour*. Sage, New York.

Schneider, B. ed. 1990, *Organizational Climate and Culture*. Jossey-Bass, San Francisco, CA.

Schneider, B., Gunnarson, S. K. & Niles-Jolly, K. 1994, "Creating the climate and culture of success", *Organizational Dynamics*, vol. 23, no. 1, pp. 17–29.

Schneider, D. 1976, "Notes toward a theory of culture", in *Meaning in Anthropology*, K. H. Basso & H. A. Selby, eds. University of New Mexico Press, Albuquerque, NM.

Scholl, R. 2003, *Organizational Culture and the Social Inducement System*. Available online HTTP: <http://www.cba.uri.edu/scholl/Notes/Culture.html> (accessed 5 August 2005).

Schneiderman, A. 2001, 'Why balanced scorecards fail!', *Journal of Strategic Performance Measurement*, January, Special Edition, p. 6.

Schneiderman, A. 2009a, Available online HTTP: <http://www.schneiderman.com/Concepts/The_First_Balanced_Scorecard/BSC_INTRO_AND_CONTENTS.htm> (accessed 29 August 2009).

Schneiderman, A. 2009b, Available online HTTP: <http://www.schneiderman.com/The_Art_of_PM/Must_a_BSC_be_balanced/must_a_BSC_be_balanced.htm> (accessed 29 August 2009).

Schwartz, H. & Davis, S. 1981, "Matching corporate culture and business strategy", *Organizational Dynamics*, vol. 10, no. 1, pp. 30–38.

Scott, W. R. 1998, *Organizations: Rational, Natural, and Open Systems*, 4th edn. Prentice Hall, Englewood Cliffs, NJ.

Scottish Centre for Facilities Management (SCFM). 2004, *Bench Marking Resources*. School of the Built Environment, Napier University, Edinburgh, UK. Available online HTTP: <http://www.sbe.napier.ac.uk/scfm/bench.htm> (accessed 5 July 2009).

Seashore, S. E. 1954, *Cohesiveness in the Industrial Work Group*. Institute for Social Research, Ann Arbor, MI.

Seddon, J. 1998, *The Vanguard Guide to Business Excellence*. Vanguard Education, Buckingham, UK.

Sekaran, U. 1992, *Research Methods for Business: A Skill-building Approach*, 2nd edn. Wiley, New York.

Selznick, P. 1957, *Leadership in Administration*. Row, Peterson, Evanston, ILL.

Senge, P. M. Systems Thinking. 1984, "Work, economics, and human values: new philosophies of productivity", *ReVision*, vol. 7, no. 2, pp. 253–1575.

Sergiovanni, T. 1984, "Leadership as cultural expression", in *Leadership and Organizational Culture*, T. J. Sergiovanni & J. E. Corbally, eds. University of Illinois Press, Urbana, ILL, pp. 105–114.

Sergiovanni, T. J. & Corbally, J. E. 1984, *Leadership and Organizational Culture*. University of Illinois Press, Urbana, ILL.

Seymour, D. & Rooke, J. 1995, "The culture of the industry and the culture of research", *Construction Management and Economics*, vol. 13, no. 6, pp. 511–523.

Sharma, S. (1996). *Applied Multivariate Techniques*. John Wiley & Sons, New York.

Shewhart, W. A. 1931, *Economic Control of Quality of Manufactured Product*. Van Nostrand, New York.

Shirazi, B., Langford, D. A. & Rowlinson, S. M. 1996, "Organisational structures in the construction industry", *Construction Management and Economics*, vol. 14, pp. 199–212.

Shivasava, P. 1985, "Integrating strategy formulation with organizational culture", *Journal of Business Strategy*, vol. 5, pp. 103–111.

Si, S. X. & Cullen, J. B. 1998, "Response categories and potential cultural bias: effects of an explicit middle point in cross-cultural surveys", *International Journal of Organizational Analysis*, vol. 6, no. 3, pp. 218–230.

Siehl, C. 1984, Cultural sleight-of-hand: the illusion of consistency. Unpublished doctoral dissertation, Stanford University, MA.

Siehl, C. & Martin, J. 1988, "Mixing qualitative and quantitative methods", in *Inside Organizations: Understanding the Human Dimension*, M. O. Jones, M. D. Moore & R. C. Synder, eds. Sage, London, pp. 79–103.

Siehl, C. & Martin, J. 1990, "Organizational culture: a key to financial performance", in *Organizational Climate and Culture*, B. Schneider, ed. Jossey-Bass, San Francisco, CA, pp. 241–281.

Silverman, D. 1985, *Qualitative Methodology and Sociology: Describing the Social World*. Gower, Brookfield, VT.

Silverzweig, S. & Allen, R. F. 1976, "Changing the corporate culture", *Sloan Management Review*, vol. 17, no. 3, pp. 33–49.

Sims, D., Fineman, S. & Gabriel, Y. 2000, *Organizing and Organizations: An Introduction*, 2nd edn. Sage, London, UK.

Singapore National Government. 1999, The Construction 21 Industry Forum: *Re-inventing Construction*. A forum jointly organised by the Ministry of Manpower and the Ministry of National Development, Singapore.

Six Sigma Partnering LLC. 2004, *What is Six Sigma?* Available online HTTP: <http://www.leansixsigmatraining.us/index.html> (accessed 5 July 2009).

Smircich, L. 1983, "Concepts of culture and organizational analysis", *Administrative Science Quarterly*, vol. 28, pp. 339–358.

Smircich, L. & Morgan, G. 1982, "Leadership: the management of meaning", *Journal of Applied Behavioral Science*, vol. 17, pp. 114–129.

Smit, I. 2001, "Assessment of cultures: a way to problem solving or a way to problematic solutions", in *The International Handbook of Organizational Culture and Climate*, G. L. Cooper, S. Cartwright & P. C. Earley, eds. John Wiley & Sons, Chichester, England., pp. 165–184.

Smith, M. 1998, "Measuring organisational effectiveness", *Management Accounting*, October, pp. 34–36.

Smith, S. 2003, *What is Culture?* Available online HTTP: <http://www.courses.fas.harvard.edu/~anth110/Lecture_Notes_and_Overheads/3_What_is_Culture/3-cultdefs.htm> (accessed 2 July 2004 compiled by Prof. S. Smith, Wittenberg University from Prof. T.C. Bestor, Harvard University).

Snow, C. C. & Hrebiniak, L. G. 1980, "Strategy, distinctive competence, and organizational performance", *Administrative Science Quarterly*, vol. 25, pp. 327–336.

Solot, M. 1986, 'Carl Sauer and cultural evolution', *Annals of the Association of American Geographers*, vol. 76, no. 4, pp. 508–520.

Sommerville, J., Stocks, R. K. & Robertson, H. W. 1999, "Cultural dynamics for quality: the polar plot model", *Total Quality Management*, vol. 10, pp. 725–732.

Spagna, G. J. 1984, "Questionnaires: which approach do you use?", *Journal of Advertising Research*, vol. 24, no. 1, pp. 67–70.

Sparrow, P. L. 2001, "Developing diagnostics for high performance organization cultures", in *The International Handbook of Organizational Culture and Climate*, G. L. Cooper, S. Cartwright & P. C. Earley, eds. John Wiley & Sons, Chichester, UK, pp. 85–106.

Stablein, R. & Nord, W. 1985, "Practical and emancipatory interests in organizational symbolism: a review and evaluation", *Journal of Management*, vol. 11, no. 2, pp. 13–28.

Starbuck, W. H. 1971, *Organizational Growth and Development*. Penguin Books, Baltimore, MD.

Steadman, L. B. 1982, "Merbs, Charles F. Kuru and cannibalism: a review article", *American Anthropologist*, vol. 84, pp. 611–627.

Steers, R. M. 1975, "Problems in the measurement of organizational effectiveness", *Administrative Science Quarterly*, vol. 20, pp. 546–558.

Steers, R. M. 1976, "When is an organization effective? A process approach to understanding effectiveness", *Organizational Dynamics*, vol. 5, pp. 50–63.

Stemler, S. 2001, "An overview of content analysis", *Practical Assessment, Research & Evaluation - A Peer Reviewed Electronic Journal (Yale University)*, vol. 7, no. 17, p. 1.

Steyaert, C. & Bouwen, R. 1994, "Group methods of organizational analysis", in *Qualitative Methods in Organizational Research : A Practical Guide*, C. Cassell & G. Symon, eds. Sage, London, UK, pp. 123–146.

Stocking, G. W. 1995, *After Tylor: British Social Anthropology, 1888–1951*. University of Wisconsin Press, WI.

Sugii, T. 1998, The construction industry suffers from declining labor productivity. An unpublished paper for the Nippon Life Insurance Research Institute, Japan.

Sutton, R. 1994, The virtues of closet qualitative research. Unpublished manuscript. Stanford University, CA.

Swailes, S. 2002, "Organizational commitment: a critique of the construct and measures", *International Journal of Management Reviews*, vol. 4, no. 2, pp. 155–178.

Swanson, G. E. 1976, "The tasks of sociology", *Science*, vol. 192, pp. 665–667.

Symon, G. & Cassell, C. 1998, *Qualitative Methods and Analysis in Organizational Research - A Practical Guide*. Sage, London.

Tagiuri, R. & Litwin, G. H. 1968, *Organizational Climate: Explorations of a Concept*. Harvard University Press, Cambridge, MA.

Tan, P. K. L. 1997, "An evaluation of TQM and the techniques for successful implementation", *Training for Quality*, vol. 5, no. 4, pp. 150–159.

Tarrant, J. J. 1976, *Drucker: The Man who Invented the Corporate Society*. Cahners Books, Boston, MA.

Tashakkori, A. & Teddlie, C. 1998, *Mixed Methodology*. Sage, Thousand Oaks. CA.

The Strategic Forum for Construction (Chaired by Sir John Egan). 2002, *Accelerating Change*. Construction Industry Council, London, UK.

Thompson, B. 1994, Why multivariate methods are usually vital in research: some basic concepts. Paper presented as a featured speaker at the biennial meeting of the Southwestern Society for Research in Human Development, Austin, TX.

Thompson, P. & McHugh, D. 1995, *Work Organizations: A Critical Introduction*. Macmillan Business, London, UK.

Thune, S. S. & House, R. J. 1970, "Where long-range planning pays off", *Business Horizons*, vol. 13, August, pp. 81–87.

Tichy, N. 1983, *Managing Strategic Change: Technical, Political and Cultural Dynamics*. Wiley, New York.

Tjosvold, D. 1987, "Controversy for learning organizational behavior", *Organizational Behavior Teaching Review*, vol. 11, no. 3, pp. 51–59.

Triandis, H. C. 1964, "Cultural influences upon cognitive processes", in *Advances in Experimental Social Psychology*, L. Berkowitz, ed. Academic Press, New York, pp. 1–48.

Triandis, H. C. 1972, *The Analysis of Subjective Culture*. John Wiley, New York.

Triandis, H. C. 1976, "Approaches towards minimizing translation", in *Translation: Applications and Research*, R. W. Brislin, ed. Wiley/Halstead, New York.

Trochim, W. M. 2001, *Research Methods Database: Measurement Validity*. Available online HTTP: http://www.socialresearchmethods.net/kb/measval.php> (originally accessed 1 June 2003 and reaccessed 5 July 2009).

Tsui, A. S. 1981, "Managerial effectiveness: a multiple constituency approach", April edn. Academy of Management, Monterey, CA.

Tsui, A. S. 1990, "A multiple constituency model of effectiveness: an empirical examination at the human resource subunit level", *Administrative Science Quarterly*, vol. 35, pp. 458–483.

Turner, B. 1990, "The rise of organizational symbolism", in *The Theory and Philosophy of Organizations: Critical Issues and New Perspectives*, J. Hassard & D. Pym, eds. Routledge, London, UK, pp. 83–96.

Tylor, E. B. 1958, *Primitive Culture*. Harper and Brothers, New York. First published in 1871.

Tylor, E. B. 1994, *Collected Works of E. B. Tylor*. Routledge/Thoemmes, New York.

University of Technology Sydney. 2002, *Benchmarking Methodology: A Tool Kit*. Webpage deleted. Previously available online HTTP: <www.pru.*uts*.edu.au/pdfs/benchreport.pdf> (originally accessed May 2002).

Urdang, L. 1968, *Random House Dictionary of the English Language*. Random House, New York.

US Department of Energy (USDOE). 2004, *Measuring Performance and Benchmarking Project Management at the Department of Energy*. National Academies Press, Washington, DC.

US General Accounting Office. 1996, *Content Analysis: A Methodology for Structuring and Analyzing Written Material*, GAO/PEMD-10.3.1st edn. US General Accounting Office, Washington, DC.

Usunier, J. C. 1998, *International and Cross-Cultural Management Research*. Sage, London.

Uttal, B. 1983, "The corporate culture vultures", *Fortune Magazine*, 17 October.

Van de Ven, A. H. & Ferry, D. L. 1980, *Measuring and Assessing Organizations*. Wiley, New York.

Van Maanen, J. 1991, "The Smile Factory: work at Disneyland", in *Reframing Organizational Culture*, P. J. Frost et al., eds. Sage, Newbury Park, CA, pp. 58–76.

Van Maanen, J. & Barley, S. 1984, "Occupational communities: culture and control in organizations", in *Research in Organizational Behavior*, L. L. Cummings & B. M. Staw, eds. JAI Press, Greenwich, CT, pp. 287–366.

Van Maanen, J. & Kunda, G. 1989, "Real feelings: emotional expression and organizational culture", in *Research in Organizational Behavior*, L. L. Cummings & B. M. Staw, eds. JAI Press, Greenwich, CT, pp. 43–103.

Varenne, H. 2003, *Culture Definitions*. Applied Anthropology Department, Columbia University. Available online HTTP: <http://varenne.tc.columbia.edu/hv/clt/and/culture_def.html#ftop> (accessed 5 July 2009).

Vesson, M. A. 1993, *Cultural Perspectives on Organizations*. Press Syndicate of the University of Cambridge, Cambridge, UK.

Vickers, S. G. 1967, *Towards a Sociology of Management*. Chapman and Hall, New York.

Walker, A. J. 1996, *Hong Kong: The Contractor's Experience*. Hong Kong University Press, Hong Kong.

Wallace, J., Hunt, J. & Richards, C. 1999, "The relationship between organisational culture, organisational climate and managerial values", *International Journal of Public Sector Management*, vol. 12, no. 7, p. 548.

Walton, R. E. 1977, "Successful strategies for diffusing work innovations", *Journal of Contemporary Business*, spring, pp. 1–22.

Walton, R. E. 1986, "From control to commitment in the workplace", *Harvard Business Review*, vol. 63, pp. 76–84.

Watt, R. J. C. 2004, *Concordance Version 3.20 - Software for Text Analysis*. Available online HTTP: <http://www.concordancesoftware.co.uk/> (accessed 5 July 2009).

Weber, M. 1947, *The Theory of Social and Economic Organization*. Free Press, London, UK. First published 1913.

Weber, R. P. 1990, *Basic Content Analysis*, 2nd edn. Sage, Newbury Park, CA.

Weick, K. 1976, "Cognitive processes in organizations", in *Research in Organizational Behavior, Vol. 1*, B. M. Staw, ed. JAI Press, Greenwich, CT, pp. 41–74.

Weick, K. 1979, *The Social Psychology of Organizing*, 2nd edn. Addison-Wesley, Reading, MA.

Weick, K. 1985, "The significance of corporate culture", in *Organizational Culture*, P. J. Frost et al., eds. Sage, Beverly Hills, CA, pp. 381–389.

Weick, K. E. 1987, "Organizing culture as a source of high reliability", *California Management Review*, vol. 29, pp. 112–127.

Weiner, N. & Mahoney, T. A. 1981, "A model of corporate performance as a function of environmental, organizational, and leadership influences", *Academy of Management Journal*, vol. 24, pp. 453–470.

Weinreich, N. K. 1996, 'A more perfect union: Integrating quantitative and qualitative methods in social marketing research' *Social Marketing Quarterly*, vol. 3, no. 1, pp. 53–58.

Wertheim, E. G. *Schools of Historical Thought and their Components by Decade*. North Eastern University College of Business Administration. Available online HTTP: http://www.scribd.com/doc/6926402/Historical-Background-of-Organizational-Behavior (accessed 5 July 2009).

Westley, F. 1992, "Vision worlds: strategic vision as social interaction", in *Advances in Strategic Management*, P. Shrivastava, A. Huff & J. Dutton, eds. JAI Press, Greenwich, CT.

Westley, F. & Mintzberg, H. 1989, "Visionary leadership and strategic management", *Strategic Management Journal*, vol. 10, pp. 17–32.

Weston, F. J. & Brigham, E. F. 1982, *Essentials of Managerial Finance*. The Dryden Press, New York.

White, B. J. 1988, Accelerating quality improvement. Presentation to the Conference Board, Total Quality Performance Conference, New York.

White, L. 1949, *The Science of Culture: A Study of Man and Civilization*. Farrar and Strauss, New York.

Whitten, P. & Hunter, D. E. 1987, *Anthropology: Contemporary Perspectives*, 5th edn. Little, Brown, Boston, MA.

Whorton, J. & Worthley, J. 1981, "A perspective on the challenge of public management: environmental paradox and organization culture", *Academy of Management Review*, vol. 6, pp. 357–361.

Whyte, W. H. 1951, *The Organization Man*. Simon & Schuster, New York.

Wilder, M. J. 2002, Exploring research designs in organizational science. Paper presented in partial fulfilment of the requirements of DBA Programme (RM502 Advanced Research Methods), Edinburgh Business School, Herriott-Watt University, UK.

Wilderom, C. P. M. & Van den Berg, P. T. 1998, Firm culture and leadership as firm performance predictors: a resource-based perspective. Centre for Economic Research. 51, 1. The Netherlands, Tilburg University.

Wilderom, C. P. M. and Van den Berg, P. T. 2004, 'Defining, measuring, and comparing organisational cultures', *Applied Psychology: An International Review*, vol. 53, no. 4, pp. 570–582.

Wilderom, C. P. M., Glunk, U. & Maslowski, R. 2000, "Organizational culture as a predictor of organizational performance", in *Handbook of Organizational Culture & Climate*, 1st edn. N. M. Ashkanasy, C. P. M. Wilderom & M.F. Peterson eds. Sage, Thousand Oaks, CA, pp. 193–210.

Wilkins, A. & Ouchi, W. G. 1983, "Efficient cultures: exploring the relationship between culture and organizational performance", *Administrative Science Quarterly*, vol. 28, pp. 468–481.

Willmott, H. 1993, "Strength is ignorance: slavery is freedom", *Journal of Management Studies*, vol. 30, pp. 515–552.

Wong, C. H. and Holt, G. D. 2001, *Contractor Classification: Multi-discriminant Analysis of UK Contractor Selection*. 1st International Postgraduate Research Conference in the Built and Human Environment, Salford University, UK, 15–16 March, pp. 280–291.

Wong, P. 2004, *Manpower for the Construction Industry: Issues and Challenges.* An unpublished address to the Master Builders Association Malaysia in 2003. Transcript not available but webpage Available online HTTP: <http://www.mbam.org.my/mbam/index.php?option=com_frontpage&Itemid=1> (accessed 5 July 2009).

Woodward, J. 1965, *Industrial Organization: Theory and Practice*, 1st edn. Oxford University Press, New York.

Wright, S. 1994, *Anthropology of Organizations.* Routledge, London.

Yau, C. 2000, "Scandals sink demand for public housing units", *Hong Kong Standard*, 24 January, p. 1.

Yuchtman, E. & Seashore, S. E. 1967, "A system resource approach to organizational effectiveness", *American Sociological Review*, vol. 32, pp. 891–903.

Zald, M. V. & Ash, R. 1966, "Social movements in organizations", *Social Forces*, vol. 77, pp. 327–341.

Zammuto, R. F. 1982, *Assessing Organizational Effectiveness: Systems Change, Adaptation, and Strategy.* State University of New York Press, Albany.

Zhang S. B. and Liu A. M. M. 2006, 'Organisational culture profiles of state-owned construction enterprises in China', *Construction Management and Economics*, vol. 24, no. 8, pp. 817–828.

Zikmund, W. G. 2003, *Business Research Methods.* Thomson (South Western), OH.

Index